FROM GREEN BERET TO SHAMANISM

ONE MAN'S JOURNEY TO HEAL

MATTHEW BUTLER

CONTENTS

"Do not judge me by my success; judge me by how many times I fell down and got back up again."
– Nelson Mandela

I would like to acknowledge my fallen brothers and sisters, those lost in combat, and especially those lost at home to combat with the demons within.

I also want to dedicate it to my family, my parents, all four of them. But especially to those who raised me and have been my role models.

I would like to acknowledge my three beautiful and amazing daughters. They are smart, independent, critical-thinking, responsible adults, for whom I am eternally grateful to their devoted mother. They accomplished this despite having a mentally ill Green Beret of a father. The odds were stacked against them, and they turned out wonderfully despite the odds.

I would like to thank the shamans Beatrice, Joan, Gabrial, Titia, and James, and the other mentors who have taught me, shared their wisdom, and have been constant and kind supporters in my shamanic journey.

Lastly, I would like to thank Sue. She was my first medicine, a wise angel, an emotionally evolved woman who teaches me every day, and she started healing me and caring for me before I knew I needed it. She gave me back my laughter and saw something I didn't see in myself.

979-8-9931996-0-3 Hardback

979-8-9931996-1-0 Paperback

979-8-9931996-2-7 Downloadable audio file

979-8-9931996-3-4 Digital online

Cover design: The Creative 5280

Interior design: Andrea Lard, The Creative 5280

Editors: Andrea Lard and Kay Lard, The Creative 5280

Printed in the United States

For questions, contact:

DoD Stamp

CLEARED AS AMENDED
For Open Publication

Aug 15, 2025

Department of Defense
OFFICE OF PREPUBLICATION AND SECURITY REVIEW

MENTAL HEALTH AND A
TRIGGER WARNING

This book discusses many sensitive topics like suicide, suicidal ideation, sexual assault, psychedelics, drugs, and alcohol.

I am writing this book to shine a glaring light on the problem that humanity is facing. More specifically, a crisis in the US and with its veterans.

The United States ranks near the bottom in terms of mental health issues compared to other countries. Mental illness is a significant issue among U.S. veterans, with **millions affected each year**. As a cohort, **7% of all veterans will experience PTSD** at some point in their lives, slightly higher than the civilian rate of 6%. The numbers are more astonishing when considering the GWOT era veterans of Iraq and Afghanistan (29% lifetime). Female veterans are more than twice as likely to experience PTSD compared to male veterans.[1]

According to the National Institute of Mental Health (NIMH), mental illnesses affect tens of millions of Americans annually.

---

1. https://www.ptsd.va.gov/understand/common/common_veterans.asp? utm_source=chatgpt.com

However, only about half of those affected receive treatment. Efforts to improve mental health care are ongoing, with research and policy initiatives aimed at better understanding and addressing these conditions, including attitudes toward mental health, our mental health providers, our mental health treatments, and our overall health care system[2].

But there's a twist. Prior to 9-11, most soldiers couldn't have articulated what PTSD was. We used to call it the "thousand-yard stare" (staring into oblivion). We had cursory classes on what it meant if someone had the thousand-yard stare, and that was about it. Then PTSD became the phrase that took center stage, and we had our wrestle with what that meant. Most of us firmly believed that any reporting of symptoms would result in loss of our security clearances and removal from the teams, the thing we loved most. So we swallowed it down. Then problems began to arise, and men, families, and units began to show cracks in their foundations. I have a clear memory of an "All Hands" meeting in JSOC where the senior NCO, CSM Thetford, called everyone in the unit (men and women from all the various classified units and the most seasoned war fighters) into a large hangar-like building where a temporary stage had been erected. He stood up there and got very real with us. For an hour, he spoke to us about his struggles with the therapy he was involved in and the false notion of losing a security clearance. He finished by encouraging each of us to seek help. This would have been in 2013, twelve years after the war began. Too little, too late.

As our collective knowledge and experience later revealed, it wasn't just PTSD, but Moral Injury. We came to understand the abundant similarities and key differences. We came to understand that a person can have both. We navigated through the difficulties of diagnosis. We struggled with Operator Syndrome. The unique circumstance that an Operator might be suffering from PTSD, Moral Injury, substance

---

2. https://www.nimh.nih.gov/health/statistics

abuse, TBIs, CEI, exposure to burn pits, and toxins, all resulting in layer after layer of similar if not identical symptoms, making diagnosis and treatment almost impossible. This was way more complex than the thousand-yard stare frame of reference we started with.

For the purposes of this book, I chose to focus on PTSD. Since this is a memoir about my journey, and since I was diagnosed with C-PTSD. However, when you read PTSD, the same short and long-term solutions apply to both PTSD and Moral Injury.

No one, not the VA, not your insurance company, not your Primary Care Provider (PCP), nor your doctor—none of them will ever care about you and your mental health or your physical health as much as you should. For that matter, Big Government, Big Business, and Big Pharma all have a vested interest in keeping you sick and depressed.

For example, one particularly bad day, after spending an hour recounting my trauma with my therapist (a proven contributor to depression), I remember riding down the elevator, and I had this vivid thought: *I am going home, feeling worse, feeling isolated, and going home to my alcoholism. Going home without any tools or coping mechanisms. He is going home to his wife, kids, boat, and vacation in the Outer Banks, and he won't think of me until our next appointment. And, if I were to kill myself, he'd simply be assigned someone else to take my slot and not even bat an eye.*

To be clear, I am not saying that these professionals aren't needed or valuable. But what I am saying is that you must be your own doctor and therapist to the single most important patient—you. After all, you know the patient better than anyone else ever will and will always have more skin in the game than they ever will. You must fight for your own mental and physical health like it is the most important priority in your life, 'cause, after all, it is—or at least it should be.

WARNINGS AND DISCLAIMERS REGARDING THE USE OF PSYCHEDELICS

While psychedelics are gaining attention for their potential therapeutic benefits, it is crucial to approach their use with extreme caution. Psychedelics can have powerful and unpredictable effects on the mind, potentially exacerbating mental health issues or leading to long-lasting psychological distress and/or permanent psychosis if certain precautions are not taken.

I encourage you to maintain traditional medicine and psychological treatments when addressing mental and physical health concerns. These established practices offer evidence-based treatments that have been rigorously tested for safety and efficacy. Working with a licensed healthcare provider, such as a licensed physician, psychiatrist, psychologist, or therapist, ensures that you receive appropriate care tailored to your specific needs.

If you are considering the use of psychedelics, it is essential to do so under the guidance of a qualified professional in a controlled, legal setting. Your mental health and well-being are paramount, and traditional methods offer the safest, most reliable path to healing and growth.

I am NOT a licensed doctor, physician, psychologist, psychiatrist, or pharmacist. I am a Seiðr and only provide a summary of techniques I have used for myself and assisting others, including naturally occurring and synthetic substances, along with well-known mental health practices.

Please be aware that I am not a lawyer, and the information provided here should not be considered legal advice. The use of psychedelics is often illegal in many states and may carry significant legal consequences. It is crucial to be fully informed about the laws in your area before considering the use of any psychedelic substances. I strongly advise you to approach this topic with legality in mind and consult with a legal professional if you have any questions or concerns.

I will only describe plants, fungi, animals, synthetics, and techniques that I have tried on myself.

Only consider the following examples after seeking your doctor's and mental health provider's advice.

UNDERSTANDING SET AND SETTING

"Set and setting" are the internal and external environments that significantly influence a person's experience with these substances. Understanding and carefully managing both can distinguish between a positive, transformative experience and a challenging or even harmful one. It is also imperative to have deliberate conversations with those who facilitate the experience.

"Set" refers to your internal state—your mindSET, mental health, mood, thoughts, emotions, expectations, intentions, and any underlying issues or anxieties you might have going into the experience. Your internal set profoundly shapes how you interpret and react to the effects of psychedelics. If you enter the experience with a positive, open, and prepared internal state, you'll likely have a meaningful and beneficial journey. Conversely, a negative or anxious internal state can lead to difficult or overwhelming experiences, often called a "bad trip."

"Setting" refers to your physical and social environment during the psychedelic experience. The setting includes the location, the people you're with, the atmosphere, and any external factors like music, lighting, or comfort levels. A safe, comfortable, and supportive setting can help you feel secure and at ease, allowing you to explore the experience without fear. Being in an unfamiliar or chaotic environment or around people who aren't supportive or trustworthy can heighten anxiety and lead to adverse outcomes.

Safety: A well-considered set and setting reduces the risk of psychological distress, panic, or dangerous behaviors. This is crucial, as psychedelics can make you more vulnerable to suggestion and emotional swings.

Intentionality: By consciously preparing your internal state and setting, you create an intentional space for exploration and healing. This intentionality can lead to more profound insights, personal growth, and positive outcomes.

Integration: The quality of the experience, influenced by set and setting, directly impacts how you integrate the lessons and insights gained. Integration is key—perhaps even more important than the actual ceremony or experience—and, sadly, often the process's most overlooked or ignored element. I follow a 1-1-1 integration process to maximize change during the 30 days of neuroplasticity. Day 1, week 1, and month 1 calls to ensure you integrate the insights gained into your life for lasting change.

Always prioritize the careful consideration of your internal state and environment when approaching psychedelic experiences. These factors are essential to ensuring that the experience is safe, meaningful, and ultimately beneficial.

The information provided here is intended solely for educational purposes. And to that end, what I have provided below is just the tip of the iceberg. There are volumes of books, podcasts, articles, and videos that dive into these topics much deeper than I can (a partial list of some of the ones I have used for research can be found in the appendix). I encourage you to review each of the modalities and suggestions I've outlined here and then conduct serious, in-depth research on your own.

A few words about Shaman, Shamanism, and Seidr and Cultural Appropriation versus Cultural Appreciation

To avoid cultural appropriation, let's talk about it.

Shaman[1]- shaman noun: sha·man | \ ˈshä-mən, ˈshā- also shə-ˈmän \ Plural: shamans

Definition of shaman: 1: a priest or priestess who uses magic to cure the sick, divine the hidden, and control events.

The term shaman[2] "comes from the Siberian word šaman. The noun is formed from the verb ša- 'to know'; thus, "the one who knows."

However, shamanism is also used more generally to describe where words like healer, religious leader, and counselor are combined. In this sense, shamans are particularly common among other Arctic peoples, American Indians, Australian Aborigines, and African tribes that retained their traditional cultures well into the 20th century.

So, what is a shaman anyway? In my personal opinion, a shaman meets the following criteria:

A shaman is not identified by a single outward sign but by a constellation of qualities that reveal their role within both the spirit world and their community. Traditionally, shamans undergo a calling, often through visions, dreams, or intense initiatory crises such as illness, trauma, or near-death experiences that break open ordinary reality and compel transformation. Being struck by lightning is an interesting indicator in some traditions. Other

---

1. https://www.merriam-webster.com/dictionary/shaman#:~:text=1,one%20who%20resembles%20a%20shaman
2. https://www.britannica.com/topic/shamanism/Selection

indicators are near-death experiences, having been born with a deformity, and having access to the three worlds (Lower, Middle, and Upper).

Unlike someone who merely studies spiritual practices, the shaman's role emerges from lived ordeal and spiritual initiation. Central to their identity is the ability to move between the human and spirit realms, entering altered states of consciousness through trance, drumming, breathwork, plant medicine, prayer, or all the above. Technique differs by tradition and location. In these states, they commune with their spirit guides, ancestors, or otherworldly beings to gather wisdom, restore the balance of energy, or perform healing.

Healing and service are at the core of shamanic practice. A shaman's authenticity is not measured by personal power alone, but by their willingness and ability to diagnose and treat spiritual causes of illness, to retrieve lost parts of the soul, to release harmful energies, or to reestablish harmony between people and nature. Their connection to the natural world runs deep: shamans are often keepers of plant medicine, animal wisdom, elemental knowledge, and sacred cycles such as the movements of the moon and stars. This knowledge is experiential and spirit-guided, not merely intellectual. Importantly, shamans are not self-proclaimed; they are recognized and affirmed by their community or elders, who witness their gifts and rely upon their service.

Ultimately, a shaman is one who embodies balance, walking between light and shadow, life and death, the human and the divine. They hold paradox, transform suffering into wisdom, and live in an ongoing relationship with the unseen world. Their authority arises from both the spirit's calling and the community's trust, making them bridge-builders whose role is to heal, guide, and restore harmony wherever it has been broken.

Many in spiritual circles suggest that the title of shaman is thrown around way too often and loosely; I agree. I endeavor to use it very reverently, with respect, and to tread lightly. I prefer to call myself a Seiðr. Seiðr is an ancient Norse magical practice, often compared to shamanism, involving prophecy, trance, and the shaping of fate. It is strongly associated with the goddess Freyja, who taught it to the gods, and with Odin, who also practiced it despite social taboos. Practitioners of seiðr could serve as healers, diviners, and spirit-workers within their communities.

Seidrs were required to enter altered states of consciousness to interact with the spirit world, where practitioners could see into the future, influence events, and communicate with spirits. This practice was often performed through rituals that included chanting, drumming, and using substances to achieve a trance-like state. Given its ability to alter fate and impact lives, it is considered a highly mystical practice.

Shamanism has been present in all cultures, in all places, and in all races for tens of thousands of years, taking on many forms and having many different words and descriptions to represent common themes and ideas. As a Seidr, my role is to help guide you on your journey into your subconscious to find the hidden roots of your spiritual, emotional, physical, and mental problems, as well as the solutions. Really, the work is done by you. You are your own healer.

And despite my best efforts to avoid cultural appropriation, there will still be those crying foul. I've learned to distinguish between *cultural appropriation* and *cultural appreciation*.

Appropriation is a seizure or capturing something for benefit, gain, or advancement—including removing it from context and using it in unintended ways and profiting from it. Examples include exchange, dominance, exploitation, and transculturation.

Here is my example: While attending a Sundance with a Lakota tribe

in Texas, I met a man who informed me about Eighth Generation[3], a

"Tribal Owned Brand" whose tag line is "Native made, not native inspired." That was the perfect encapsulation of the differences between Cultural Appropriation and Cultural Appreciation. If you walk into a high-end boutique in Park City, Utah, and buy a turquoise-encrusted Buffalo Skull sold at $10,000, and zero of that money reaches a native tribe, artist, or person, this is cultural appropriation. However, if a person were to shop for a similar piece directly from a native artist so that all the proceeds go directly to the artist, their family, and their tribe, well, that is Cultural appreciation.

ABOUT MY FIRST RETREAT
AND CEREMONIES

Anyone familiar with ayahuasca ceremonies will quickly point out that much of what happened is highly outside of norms. Some may even question the safety of what happened, although I never have and never will. I'd like to suggest that these were extraordinary circumstances for an extraordinary necessity. One of my many mentors also taught me, "Every ceremony is perfect." If we only understood how deeply the Divine is involved in orchestrating every ceremony, we would realize how deep that lesson runs and how literal it is that every ceremony is perfect when done in love, light, and partnership with the Divine.

ABOUT MY COMBAT EXPERIENCE

Upfront, I want to set the record straight. I want to walk the fine line of complete honesty, neither embellishing *nor* minimizing the events that brought me here. I was diagnosed with Complex PTSD (C-PTSD) in 2011. It isn't all that surprising that a Green Beret with 40 months in combat and spanning six deployments in both Iraq and Afghanistan would have C-PTSD, but it isn't entirely due to war. My C-PTSD came from several events in my life, not just combat.

I have experienced firefights, mortars, rockets (very few), and the exterior wall of our compound being breached at 0400 in the morning, 50 meters from where I slept. I watched those I led in combat having their plane hit with a rocket moments after I put them on that plane to go home (they all lived)—ironically giving up my seat in what I thought was a selfless act.

Psychologically, I experienced loss, mourning, survivor's guilt, and the constant worry of unpredictable death, fear, and long hours—days in and days out for months and months on end. I felt loss, separation, and grief, and I had the sacred role of holding a man in my arms as he died.

In my civilian life, I experienced abandonment as an infant, adoption, body shame, bullying, sexual assault, three divorces resulting in two short stints of homelessness, and excommunication from a rigorous religious community.

All played a part in my mental illness.

Formally, I was diagnosed in 2011 with Traumatic Brain Injuries (TBIs), addiction, alcoholism, C-PTSD, treatment-resistant depression, and suicidal ideation.

With respect to my combat, I spent most of my deployments as a staff officer—as an Army officer—the typical career path that takes you out of direct combat. I only experienced a tiny sliver of the intense combat that most of my Special Forces brothers did. I did not have the fierce firefights or the prolonged, sustained combat that many others have had, especially those Non-Commissioned Officers (NCOs), the real heroes of the War on Terror, the men who day in and day out for twenty years went into harm's way countless times. We all collectively owe them a debt of gratitude that we will never be able to repay.

PREFACE

Dear Reader,

Who did I write this book for? I wrote it for the "lost" masculine. If you are a man and feeling lost or struggling with addiction, alcoholism, depression, or suicidal ideation, or wondering if there is any meaning or purpose to this life, this book is for you. Or if you know a man who meets the criteria above, this book is for those you love and are hoping and praying they will return. As a retired Green Beret who deployed to Afghanistan and has a total of 40 months in combat, I have learned a few lessons along the way. I wrote what I learned in my healing journey to come home to myself and share it with whoever will listen.

This is the story of my past lives, trauma received, trauma inflicted, and ancestral trauma inherited. This is the story of my brothers and sisters who fought beside me, before and after me. This is a story of plant medicine, holistic healing, and mental health.

This is not a war story, although it is a story told in the context of a war. This story is about a new war, a war on corrupt governments, self-serving politicians, and a criminal justice system that lacks justice.

This is the story of everyone who screams for a solution and intuitively knows there is one out there and just can't find it.

This story is about shamanism, an ancient healing practice nearly 100,000 years old. This is a story about love and forgiveness, a story about redemption and healing.

It is a love story, a story of learning to love and forgive myself.

But most of all, it is a true story.

I hope you enjoy the book.

PART ONE
THE GREEN BERETS

1

WHAT ARE THE ODDS?

"I was robbed of the person I was supposed to be. I don't fit in anywhere, not with my adoptive family, not with my biological family. I'm like a puzzle piece that was cut apart to fit into a puzzle it didn't belong to... The damage was done. That's what adoption has done to me." - Jewel Kingsley

As my birth mother would describe it almost 45 years later, she held me in her arms and stared into my eyes, trying to memorize my face, my smell, and my fingers gripping hers. She was convinced this would be the last time she would ever see me.

My biological mother and I share many similarities. She is of Scandinavian descent, Southern Sweden and Northern Denmark to be exact. This information would later come to serve as a breadcrumb[1] in my journey to discover who I am. She, too, was

---

1. For context- I use the term "breadcrumb" to illustrate the process I was following

raised in a small town in Northern Utah with a strict Mormon culture and home. It's a culture built around fear and shame. When she put me up for adoption, I would be placed in a Mormon home and end up living in a small Mormon town.

She later admitted that she had a lifelong struggle with body issues and struggled with her self-worth because of that. She was intelligent and excelled at academics and had been awarded a scholarship to Brigham Young University (BYU)—the school reserved for the most devout of the Mormon faithful and, notably, voted the Most Cold Sober University in the country. My mother wanted to use her gifts to become a nurse.

In Alcoholics Anonymous (AA), I learned of something they call the God-sized hole. It's a hole that I suspect most of us develop in our lifetimes, a hole that, no matter what we use to fill it, never gets filled until we learn to fill it with God or a higher power. Well, my mother had a God-sized hole.

She was in her junior year when she felt the need to self-medicate that wound or fill her emptiness, her inadequacy, or her lack of self-esteem. That's another trait I shared with my biological mother. So, one Autumn night, she sought the intimacy of a stranger at a party at a neighboring university where alcohol was more free-flowing, to find self-worth, to feel valued. It was her first time, and that was all it took; she was pregnant. Unknown to her, my biological father, who was of Germanic descent (another breadcrumb), was married and promptly abandoned us.

This fact left her to choose from three of four probable choices. All women in this situation have four choices—none great: 1) Marry the father and try to create a family. Sometimes, this works out, but most of the time, it doesn't, which wasn't an option for her. 2) Abortion.

---

after my spiritual awakening to discover deeper truths about myself, such as who I really am.

She said that she couldn't bring herself to do that. 3) Raise me as a single mother spanning the 1960s, 1970s, and 1980s—a time in Utah that wasn't particularly hospitable to single mothers in staunch Mormon towns and cities. She knew that she was going to be unable to do this. She knew that she would struggle, that we would struggle, and that there would be untold hardship for both of us. 4) Adoption. She worked with the Mormon church to have me placed through their welfare system with another Mormon family. She figured that at least I'd be brought up with a solid religious foundation despite her having become an ardent atheist by this point.

She dropped out of BYU for a year and moved to Park City, Utah, the only liberal-leaning town in the state then, where she could fly under the radar. She moved in with an older, married sister, told the prying neighbors a lie about where her nonexistent husband was, delivered me, then promptly vanished and returned to BYU to ultimately become a psych nurse.

Ironically, my biological father had gotten both my mother and his wife pregnant at what must have been within a few days of each other. I have only had three or four conversations with my biological father; he doesn't seem particularly interested in maintaining a relationship. But during our first conversation, he asked, "When were you born?"

I replied, "June 25th, 1969."

"Wow," he exclaimed, "your half-brother Steve was born on June 24th, 1969."

"Interesting," I said as I began to do some math in my head.

"Where were you born?" he asked next.

"Just a moment," I said as I sifted through the documents placed on the table before me to answer and verify my claims, should he call our relationship into question.

"Here it is," I said, picking up my birth certificate. "It says I was born in St Mark's hospital."

"Wow," he exclaimed again, this time with more intensity, "That's the same hospital that Steve was born in."

"What are the odds?" I quipped. In the back of my mind, my real thought was, "WTF?"

That's right, my mother was sequestered to her room because of the shame of an unwed, unplanned pregnancy, which was the protocol in those days. At the same time, my father's wife was a few doors down, on the same ward. They kept you in a maternity ward for three days after giving birth back then. And yes, my half-brother and I lay in our respective bassinettes in the same nursery.

His name was Steve. Steve was the inspiration for the character Heroin Bob in the movie *SLC Punk!*, although the details of his life and the character's fate in the film diverge from reality. Steve was a friend of director James Merendino during his youth in Salt Lake City. Although the character in the film is fictionalized, many of Heroin Bob's traits and experiences are inspired by Steve, particularly his ironic anti-drug stance despite being part of a scene where substance use was prevalent. In interviews, Merendino mentioned that *SLC Punk!* was a semi-autobiographical film, drawing heavily from his own experiences and the people he knew in the punk scene of Salt Lake City during the 1980s. Unironically and sadly, Steve would later die from a heroin overdose in 2006, about 6 years before I reconnected with my biological family. So, as Dad gazed into the nursery at his son, he was looking at his two sons. It was a hell of a way to start my life's journey.

2

A ONE-HORSE TOWN

"I drove a car at 15 while my mom was at work, they're still talking about what a hoodlum I am. I am now 73."- Anonymous

My adoptive Mom and Dad are wonderful; I love them more than words can express. They went through a lot to raise me; I was not easy by any standard. They were and are exactly what my bio-mom had hoped for—a solid home, traditional family, and strong Mormon culture and teachings. My father attended the University of Utah, majoring in business. He found work in banking and would shortly be offered an opportunity to manage a bank in the rural town of Coalville, Utah.

When I was two, we moved to the neighboring town of Henefer, population 600. In many ways, it was an ideal childhood. I had safety, protection, a tight-knit community, close friends, and freedom to roam the outdoors. This was the blessing of growing up in Henefer. There were mountains, lakes, rivers, and wildlife. Almost all my free time was spent outdoors. I learned to hunt, fish, and navigate with a map and by stars.

One of my fondest memories is of my dad teaching me how to navigate by the North Star, which would be another breadcrumb. I learned to ice skate on frozen fields, played every sport you could imagine, and learned how to work hard—not just ordinary hard work but ranch work—at an early age. In the summers, my mom would wake me up at 5:30 to ensure I had breakfast before I was sent off to the hay fields at 6:00. When people ask me if it was hard to become a Green Beret, I always say yes because it was. But this voice in my head always reminds me that I'd been a Green Beret in training for as long as I can remember. There was very little the Army had to teach me that I hadn't already learned when I was a child; they just knocked the rough edges off. In the book *The Company They Keep*, Anna Simons explores many personality traits that make individuals more likely to succeed in Special Forces. Common traits include:

- **Rural Upbringing**: Individuals from rural areas, who tend to be more self-reliant and familiar with physical labor, are often well-suited for the challenges of Special Forces training.
- **Participation in Combat Sports**: Many Green Beret candidates have backgrounds in physical, combative sports like wrestling or boxing, which helps build the toughness and endurance needed for Special Forces.
- **Broken Family Homes**: Broken families sometimes develop an early sense of independence and emotional resilience, critical traits in Special Forces operations.
- **High Intelligence but Poor School Performance**: Many successful Special Forces soldiers are highly intelligent, but their school performance may not reflect their intellectual abilities due to non-conformist tendencies or boredom with traditional education.

I was four for four if adoption counts for a broken home.

Our house was a 100-year-old farmhouse. It was drafty and cold in a town that experienced several days below zero each winter. Our primary means of heat was a large wood stove that sat in the center of the home. We would spend our summer weekends loading the truck with wood logs, cutting them to 18-inch lengths, and then splitting them with an ax.

Even as a young kid, I was becoming acquainted with rage. My father killed two birds with one stone. Whenever I would show my rage, he would send me to the woodpile. I'd be told how much wood I had to chop before I could go play or spend time with my friends.

He always said, "Son, you better control your temper, or you are gonna wind up in jail someday."

Dad would later be the one who called the cops, fulfilling his own prophecy. By the time I was 12, it was my job to wake up, feed and water the hogs, sheep, horses, and chickens, and bring in wood to heat the home before I could eat breakfast. I then had to be at the bus stop by 7:00; we were the first stop in town. Then, ride the bus for 45 minutes to school.

My childhood was idealistic in some ways, and in others, it was problematic. Small towns with a population of 600 are very close-knit. There is no room for error in a small town. When an adult is your football coach, best friend's dad, girlfriend's uncle, religious layperson, and high school science teacher, it presents certain problems. Specifically, if you make a mistake in any one area of your life, it suddenly impacts every area of your life; nothing is compartmentalized.

There were four prominent families in our town. If your surname wasn't one of those four names, you were an outsider, and your fate was sealed. It wasn't very hospitable for outsiders. Despite living every day of my life from 2 years old to 18, I was and am still considered an outsider. As recently as a few years ago, my father was explaining to my oldest daughter that, despite having served multiple terms on the town board, multiple terms as Mayor, Bishop of the local Mormon congregation, and having lived there continuously for over 50 years, the town still sees him as an outsider.

Yes, I had a handful of close friends, who I am still close to today, but most of my 50 classmates would reject me. I would go on to spend the remainder of my childhood struggling with my identity and trying to find my place in that world. I was a small fish in a very small pond filled with fish who seemed larger than life, cowboys, and such. In a town where EVERYONE, and I mean everyone, had a nickname given to them, mine became Bubba. To this day, when I go to Henefer, everyone will call me Bubba, unaware of the pain that this name has caused me over the years.

I would later learn that trauma is what often kick-starts a person's spiritual gifts. Trauma frequently causes dissociation, and shamans deliberately disassociate with intention, a way to bridge the two realms.

When we moved to Henefer, our first home was in an apartment on Main Street. In those days, the town's streets were lined with open irrigation ditches. These ditches were about two feet deep and three feet wide. They were filled to the bank with swift-running water. One day, while playing in the front yard, I fell into the ditch. I was pulled out several minutes later by an older neighbor. This would be the first of many close calls with death. Later, while studying shamanism, I learned that Near-Death Experiences (NDEs) were common hallmarks of shamans or people who had gifts. This crossing over had the effect of thinning the veil, as it were. As I continued to grow and become more conscious of events, I became aware that I was having unexplainable experiences. I would often feel like I could feel the vibration of objects, or talk to plants and animals. I even saw ghosts regularly in the farmhouse where I was raised.

One morning, after coming out of my body (astral projection, another shamanic technique) the night before, I asked my dad about these experiences and how to manage them. His response was to label them as "occult and satanic," and he made me promise that I would never do it again. This was a key chapter in my development; it would later become a significant breadcrumb. Wanting my father's approval, I buried my gifts so deeply that I forgot about them altogether.

When I was eight years old, I woke up with an upset stomach. I complained enough that I was given the standard sick child protocol and was handed a bucket and a blanket and sequestered to the couch out of view of the TV. My mom had arranged a doctor's appointment in the afternoon, assuming it was the flu. By noon, I was goofing off and acting normal. My mother thought I was faking it and debated whether or not to take me to the doctor's appointment. Ultimately, she decided to take me. After waiting in the waiting room, I was given a routine exam.

"Mrs. Butler, we are rushing your son to Cottonwood Hospital." Doctor Snarr said.

"What?" my mom asked in shock.

"Your son's appendix is about to burst, and we need to perform an emergency appendectomy right now." He informed my mother.

I was told to wait while an ambulance came, and my mother followed us to the hospital. Again, I narrowly escaped death.

At the age of twelve, I went to work at the local truck stop washing dishes. This job was an informal education in the ways of the world. As it turned out, there was one small crack in the walls that was meant to keep the outside world out—the truck stop at the interstate exchange between I-80 and I-84. It was called the Kozy Café and was a real landmark for truckers. This was before the days of national chains, and truckers stopped there religiously. I would learn about cigarette smoking, beer, "lot lizards," and 70s country music. One of my so-called mentors was a man named Patrick.

Patrick always sat at the back table in the far corner. Patrick was old but looked much older than his real age. The coffee, whiskey, and constant exposure to the weather due to his profession as a sheep herder—not to mention his chain smoking—had left his skin dark and wrinkled. Patrick taught me a lot about being a man during our informal chats while I bused tables, as he sipped coffee waiting for his wife to finish her shift. To this day, my heart skips a beat if I hear a Conway Twitty, Loretta Lynn, or Kenny Rogers song. It was a formative time, a time when I began to explore my surroundings and engage with the outside world.

There was another dishwasher there, one of the popular kids. He liked to tease me and bully me as often as he could. One day, we got into another of our many fistfights when he was picking on the obvious: my weight. We fought until I was kneeling over him, hitting him in the nose until the blood ran profusely down his cheeks. I'd stop the pounding long enough to ask, "Have you had enough?"

"Yes," he would choke out through his blood and tears.

I would get up to leave and walk away, and hear some taunt from behind me as if he thought I wouldn't turn around. We would re-engage in the struggle, winding up back in the same scenario again. This repeated three times until I finally broke his arm. That day, during my shift, Patrick confronted me.

"I heard you fought with Trevor today at school?" Patrick asked.

"Yes, sir," I replied, emptying ashtrays into my busing tub.

"I heard you broke his arm." Patrick continued.

At this point, I began to get concerned. I assumed that as a local, Patrick would side with Trevor and his family.

"Yes, sir," I replied again with a knot in my stomach.

"Good, that little shit deserves it." Patrick said with a calm air of approval, "When you turn 21, remind me, I'll buy you a beer."

Patrick died before I collected that beer. Trevor never bullied me again; that's how you handle things in a small town.

During high school, I participated in Future Farmers of America (FFA), 4-H, football, wrestling, track, and rodeo (Our school had a rodeo team and a ski team). But if I had one sport that I loved, it was wrestling. I loved the one-on-one combat. Our coach always said it was the most difficult sport because of the solitary nature of one-on-one combat. But the truth was, I sucked at wrestling.

Wrestlers typically wrestle 24 matches a season. In my first two years of high school, going into my last match of the year, my record was 0-47. I had lost 47 straight matches in two years. But something about growing up in the small town, wanting to succeed at the toughest sports available, and wanting approval made it so I would never quit. This character trait would serve me well on my path to becoming a Green Beret and shaman. I won my last match to finish out the two junior Varsity years at 1-47.

My next brush with death happened when my best friend and I were in a car rollover during the summer between my sophomore and junior years. It happened after a late-night shift at the local resort where rich city folk came to take advantage of the mountain lakes. We both fell asleep just a few miles from the resort, and the car careened off the road, making an in-air barrel roll, *Duke's of Hazard* style. I was thrown out and forward of the spinning car, which landed and then rolled over me. The next thing I remember was waking up in my parents' living room at 5:00 a.m.

At the same time, puberty came on in a big way. I knew I was bigger; hell, I outgrew all my clothes, but I had no idea how big I was. When my junior year of wrestling started, the coach asked me what weight class I wanted to wrestle. My last weight class as a sophomore was

145 lbs., so I answered with "167," two weight classes above 145. I stood on the scale and was easily over 167.

I said, "187, I guess?" Responding to my failure to meet the 167 weight, I bumped up to the next higher weight.

Again, I was over that. Coach responded, "You'll wrestle heavyweight," and waved me to get off his scale. I had gained 6 inches and 50 lbs. in one summer. I wrestled as the varsity heavyweight for the next two years, placing 2nd in regionals and 6th at state in my final year. Not a state champion, but a reward for not quitting in my first two years.

I credit my unwillingness to quit to my father. My father always enforced the rule that if we started something, we finished it. It helped to correct impulsive decisions. Finishing is the same as quitting—once either of them becomes a habit. My father wisely helped me form the habit of never quitting. It would become one of my most valued habits and something I could rely on throughout my life.

Growing up on a farm also taught me the value of hard work. Hard work doesn't require talent. Later in life, when I went through the multiple selections of the military, I would surmise that I was rarely the smartest, fastest, or strongest soldier in the class. But I could always count on being the hardest worker in the class. I knew I would be fine if I were willing to show up early, stay late, and do the work.

Another value my parents taught me by example was selfless service. For all the many things that the Mormon church is, it does operate on the principle of service. Living three miles from two major interstates meant that when someone broke down and needed gas or food, the locals directed them to our house since my father was the lay clergy. In all my years, I never saw my father turn anyone away, a living example of the "giving the shirt off his back" type of man.

3

A BRAVE NEW WORLD

"The only way to make sense out of change is to plunge into it, move with it, and join the dance."- Alan Watts

The summer following my graduation from high school, I worked as a landscaper in Park City. We country kids referred to this part of the county as "Posh." It is where all the ski resorts and the Olympics were held in 2002. There, I met a young man named Leo, who would later become my college roommate. He was originally from South Carolina, and following his graduation, his parents moved back. I asked my parents if we could take him in until college began a few months later. They agreed, and later, he would ask me, "What are you doing this fall? Where are you going to college?"

I hadn't decided if I was going to college, let alone where. I had applied to only one university, the University of Utah, and was denied. I struggled with deciding to join the Army or attend college and serve a Mormon mission. What I really wanted was to enlist in the Army as an Infantryman. I came from a military family. My father was an infantryman during the Korean Conflict, and my uncle was in Vietnam. My grandpa was a sailor in the Pacific theater during WWII, with others in WWI, the Civil War, and the Revolutionary War.

But then, I was only 17 and would need my parents to sign my enlistment papers. At that time, one of the questions was, "Had I ever used marijuana?" As a young teen, I had, on a couple of occasions, smoked a few joints, nothing much to speak of. Still, I felt like the Army was some sort of all-knowing entity that couldn't be lied to, so I was honest on the papers. I also didn't want my staunchly religious parents to discover my "drug" use, so I declined to join the Army and had no plans.

On the horizon was a Mormon mission. A Mormon mission is an obligation for young men. At that time, you entered the "mission field" at 19. I had always been expected to go, but I wasn't sure if I wanted to. However, the guilt and shame dished out by my family and the entire community made it nearly impossible to escape. A few older friends had chosen not to go, and their lives were a living hell. I would go out of fear of disappointing my parents and to avoid the shame of not serving.

Leo asked me to be his roommate at Weber State University, and I thought this was a good choice since I had a year before I would serve my mission. Our first year of college was exciting and had little to do with academics. We met a couple of others from our floor in the dorm and had a tight circle real fast. We explored fraternities, girls, alcohol, and freedom. Of the four of us in that group, Leo would go on to join the Marines, Adam the Army, Donny became a SWAT sniper, and I would serve a mission before I would ultimately join the Army. I was called to serve in the Ireland-Dublin Mission; it would become another breadcrumb.

I loved Ireland and the Irish people, serving a mission—not so much. Daily rejection, long hours, monastic living conditions, and strict rules. Despite these conditions, I worked hard as a young Utah farm boy would. I had my successes and made great friends. I came to love the Irish people and culture a great deal. I was fortunate to see nearly every one of the counties, and most major cities. I lived in Belfast, Dublin, Sligo, Limerick, and Dublin again. I would also visit other Neolithic sites like the Hill of Tara. Unbeknownst to me, I was following more breadcrumbs that I would later revisit.

My aunt had traced our family history and the Butler name back to County Kilkenny. In ways, I would identify more with my ancestors and my Irish heritage than my Utah Mormon identity. My adoptive mother is English, and my adoptive father is Irish. Genetically, I am 45% Scandinavian (Biological Mother), 45% German (biological father), 5% Iberian, and 5% other. I believe that my most recent ancestors came from Ireland, and they were preceded by the Scandinavian and Germanic tribes that fought and intermingled as they conquered their way through Europe, namely England, Scotland, and Ireland. This will go on to be an important distinction in my journey.

Belfast was divided into three areas. One affluent, one Catholic, and one Protestant. Think of it as three slices of pie. The protestant area was to the west, the catholic area in the middle, and the affluent area, where we lived, to the east. The only way to travel from the east to the west was to ride our bikes into the city center and all the way into the heart of the area. The roads were maintained by heavily armed British soldiers in armored personnel carriers. The city was filled with remnants of bombed-out buildings, burned cars, wire strung across the streets, and broken glass on the top of the walls. Many of the men we met had been "kneecapped" (shot in the knee so that their legs would have to fuse to heal). This was another great lesson about affluence and poverty. When there was plenty, no one fought. No one had time for the trouble they knew that fighting cost.

In this area, Protestants were married to Catholics. People were less dogmatic in their beliefs. Protestants lived next to Catholics, and nobody cared. But in the slums, everyone cared. Everyone was embroiled in the fighting. We had a curfew of 9 p.m., but one day, we were teaching, and it ran long. We found ourselves out past curfew and decided to take a shortcut home by riding straight to our residence rather than making the long trek into the downtown area and back up to our flat. We would be coming straight out of the protestant area, crossing the heavily patrolled Shankill Road, cutting into the catholic area, and continuing into the affluent area. Only minutes into the catholic area, we found ourselves being followed by six young men ages 16-18. No matter what we tried, we couldn't evade them. Ultimately, they cornered us, each holding bottles, rocks, or sticks, ready to defend their homes from outsiders.

"Would ye be Protestant or Catholics?" the leader asked us.

"We're Mormons," I replied, stone cold.

"Would ye be Protestant Mormons or Catholic Mormons?" he probed deeper; he was no theologian.

And out of nowhere, I heard a voice in my head and responded with, "We are Jewish Mormons." I guess two can play his game, I thought.

He stared deeply into my eyes, waiting for us to blink or flinch, and then replied, "All right then, ye can pass."

And we rode our bikes home as fast as we could and swore never to try that shortcut again.

One day, while serving in Dublin, I got a letter from my parents. It was filled with newspaper clippings from Utah newspapers. The headlines all said the same thing: "Two missionaries serving in Bolivia are assassinated by terrorists." My childhood friend Jeffery Ball was murdered by terrorists while serving on his mission in Bolivia on May 24, 1989. The day he was murdered was his "P-Day" or preparation day. On P-Day, you spend half the day shopping, doing laundry, and writing letters home, and the other half doing service projects. He

44

had spent the last half of the day working at an orphanage and was shot on his doorstep.

His needless death sent me into a depression cycle, and I didn't leave our apartment for a week. I began to question the existence of God and missionary work. I had no idea at that time just how intertwined my life would become with terrorism. A few days later, I received a postcard from Jeff that he had mailed the day he was murdered. He had written, "I never said it would be easy; I only said it would be worth it."

In those days, many Dublin homes were heated with coal. One day, while knocking on doors, my companion and I were working on the same street where the coal provider was making deliveries. There were four of them following a flatbed truck, unloading 50lb bags of coal at the customers' doors. These large, soot-covered men naturally took to mocking us and making derogatory remarks as we worked our way down the street.

The lessons of my youth were still fresh: you deal with a bully head-on. I handed my bag and coat to my companion, turned, and started walking directly toward these four men. My companion asked me where I was going, but I didn't turn around to answer.

I approached the truck and looked at the man offloading the coal. I asked, "What door?" and held out my hands.

He laughed.

"What door?" I asked again, this time with my eyes more determined. The others cheered, encouraging him to hand me a bag of coal and break my spirit. He didn't know that I had spent nearly every day of every summer hauling 50lb bales of hay. I flung the coal over my shoulder, dropped it on the designated doorstep, and returned to my place in line for another. This time, I asked for two. Oddly, their bullying jeers fell silent. Now, I was carrying 100 lbs. of coal in a white cotton shirt and dress pants. I kept up with them for the remainder of the street. After that day, whenever we crossed paths with them, they would tip their hats and nod at us.

Toward the end of my mission, I served as the Assistant to the President or AP. The APs work directly for the Mission President, a middle-aged or older man who has retired and is there to provide some degree of adult supervision and liaison with local government and civic organizations.

One night, the Mission President called me, asking, "Do you know where 'such and such' hospital is?"

"Yes," Was my quick reply.

"Can you drive me there? One of our missionaries, Elder Critchfield, is in the hospital," He stated.

Elder Critchfield had been a flatmate of mine for a few months, only recently moving to a new area that I had been part of assigning him to.

"Sure, I'll meet you at the office in five minutes."

My companion and I walked the short distance to the mission home office and waited for the mission president to pull up in his car. I jumped into the driver's seat. We drove to the hospital, where we met a doctor who asked us if we were connected to Elder Critchfield.

"Yes," the president replied.

"I regret to inform you that he has passed away."

Gale Stanley Critchfield, 20, who had been in the mission field for 15 months, was attacked near the front door of his apartment as he and his companion were walking home. He died in a hospital the following day of a stab wound to the heart on May 27th, 1990, one year and three days following Jeff's death. I must have been the only person in the world to know both murdered missionaries. It seemed like the universe was giving me lessons in death, suffering, and grief at a very early age. I was growing up quickly. When other young men my age were tailgating at college football games, partying, and experimenting with all life had to offer, I was mourning my friends.

I would end my mission a few weeks later, in early July 1990, and return home. One of the first things I did was call a US Army recruiter. I was now over the age of 21 and an adult. I wouldn't need my parents' permission, and I wouldn't disappoint them for being a "drug user." If only they knew what the future would hold—

I joined the Army Reserves on August 1st, 1990, as a 25S (combat photographer). My lack of familiarity with the Army and a desire to stay in the Reserves and continue studying at WSU, as well as my years on the high school Yearbook committee, meant that I already had considerable experience with photography. All these factors heavily influenced my decision to choose 25S. One day later, on August 2nd, 1990, Saddam Hussein crossed over into Kuwait. This would come into play several more times throughout my life.

I completed one more college semester while waiting to ship out to basic training. I attended basic training from January to March 1991 at Fort Leonard Wood, Missouri. Our shortened sleep cycle was interrupted each night by an additional fire watch duty. Fire watch was how the Army ensured its dry, wooden WWII barracks wouldn't burn down. During the Gulf War, our Company instituted a second fire watch. The additional fire watch would be responsible for watching CNN and making a report to inform us of the war's progress. As the US began to gear up and pre-position soldiers and equipment in the Persian Gulf, the Army's newest soldiers were filled with a strange mix of bloodlust and fear.

We were confident that the war would go on for years and that we would deploy as soon as our training was over. We spent late nights cleaning our rifles and would talk into the night about how soon it would be before we would find ourselves in combat. Looking back as a 27-year veteran, this idea elicits a chuckle. There was almost zero chance of an Army reserve photographer seeing any action immediately following initial training. Still, I was 100% confident of it in those days as I rehearsed tossing grenades, stabbing rubber dummies with my bayonet, and shooting human silhouette targets at distances of 300 meters.

Basic training was easy. Again, I credit it to having been raised in a hard town by rugged men, doing real work for over a decade at this point in my life. The war ended in 100 hours, barely longer than a holiday weekend. I felt like I missed the opportunity of a lifetime, a chance for battle.

Following basic training, I was sent to what was once Lowery Air Force Base in Denver, Colorado, for my Advanced Individual Training (AIT). It was six months, which is exceptionally long compared to the other job skills in the Army. It was also considered relatively cushy compared to my basic training cohorts, who were still stuck in the Missouri Ozarks in March. We enjoyed meals in a cafeteria and were allowed to sit down and eat. No one shouted at us to eat as we walked from the end of the line to the exit. We had free time and the freedom to go off base. This was a strange new world. Having come fresh off my mission, I was still very devout and doing my best to follow my Mormon religion. I found out that there was a building with the local congregations only a few miles away. I mastered the bus system and attended a Sunday meeting shortly after arriving in Colorado.

Mormon culture is heavily centered on the family and abstaining from sex outside of marriage. This leads to marriage at an early age for a large majority of traditional Mormon faithful. To facilitate this, the Mormon leadership created what are known as "singles wards." The sole purpose is to gather the young Mormons into a single congregation, watched over by a married Bishop, to encourage this early marriage dynamic. Imagine my surprise when the congregation closest to my base was a singles ward.

It worked; I met my first wife, Sophie, shortly after. We became close friends because she had a boyfriend serving in the Air Force who was stationed in the Azores Islands for a year. They planned to marry when the year was up, and he returned. She was my best friend at this time in my life, and I respected their relationship. I began to date other women in the singles ward. Shortly after one of these dates, on Memorial Day weekend, Sophie saw me with my date while we were at a party, and an expression of despair came across her face. The rest of the evening was awkward as I tried to be present with my date while knowing that Sophie was having feelings for me.

I also had feelings for her, but I hadn't acted on them for the reasons I mentioned. But now things have changed. We spent the evening talking in the parking lot of my barracks and mutually decided to start dating. Five weeks later, on July 4th weekend, I proposed, and she accepted. We set the date for August 31st and had a whirlwind courtship.

For those unfamiliar with Mormon culture, this can sound very shocking, as it should. But for anyone in the Mormon community, this is totally normal behavior. I sincerely loved her; she is such a wonderful woman, and we had all the love, euphoria, romance, and growth that courtship can provide in three short months. Additionally, we had our faith in common, and there is something to be said for faith and a strong moral context that gave a framework for building a family and a life together. We were married in August 1991.

Following a honeymoon to the Pacific Northwest, we returned to Utah, where I remained in the Army reserves, attended and worked at WSU, and took Reserve Officer Training Corps (ROTC) classes. Three months later, Sophie was pregnant. It was a conscious decision —another Mormon influence. We were excited to be parents. Our first daughter was born in August 1992. I graduated from WSU with a bachelor's degree in history and a minor in Military Science

(ROTC). I chose history because, at the time, my focus was to go on active duty or teach history.

As a student teacher, I quickly learned that I loved teaching but hated teaching teenagers. I soon abandoned any possibility of becoming a teacher. When I graduated in June 1993, Sophie was three months pregnant with our second child. Simultaneously with graduation, I was commissioned as a Second Lieutenant in the Utah Army National Guard Field Artillery. Although I had been number one on our ROTC Order of Merit List (OML) and would have been guaranteed an active-duty commission, my Branch of choice, and duty location of choice, I chose the National Guard. I decided to go into the National Guard because Sophie, while very patriotic, was afraid of the possibility of me going to war and being killed or wounded and being widowed.

Soon, we would leave for Fort Sill, Oklahoma, to complete my Officer Advanced Course (FA OBC). We were stationed at Fort Sill from September 1993 to April 1994. Our second daughter, Heather, was born in January 1994. While there, I did my Physical Training (PT) each morning. I was secretly training for Ranger school. Ranger school is the premier Army leadership school, customarily attended by those in the combat arms or assigned to the Ranger Battalion. My PT score was the highest in the class, scoring 310 points on a scale of 300. I hoped the National Guard would send me to Ranger school, but those hopes would be dashed.

With a history degree, unwilling to teach, and having turned down an active-duty appointment, we returned to Ogden, Utah shortly after, where I would continue to serve in the National Guard and sell real estate. Selling real estate was soul-crushing. I was good at it and earned the distinction of "Rookie of the Year" in my first full year at the job. I tried very hard to sell ethically and honestly. I felt that my clients had a good experience and that I took good care of them. However, I repeatedly found that my colleagues had very little professional loyalty. I was dying inside despite success. On top of that, real estate was a financial roller coaster, either feast or famine, nothing in between. During this time, our third and final child, Raquel, was born in September 1996.

One of my ROTC mentors invited me to run my first marathon to stay fit and military-minded. This mentor was a Special Forces officer in the 19th Special Forces Group and had nearly qualified for the Winter Olympics in biathlon. I was no match for athletic prowess, but I was a stubborn farm kid who didn't know how or when to quit. We trained and completed multiple triathlons leading up to the St. George marathon, which he was using to qualify for the Boston Marathon. This meant that he would need a sub-3:00-hour marathon. Me—I was just out to finish.

Less than a week before the marathon, we did a light-paced eight-mile run in the mountains east of town. I got sand in my shoe, and like a stubborn farm kid, I kept running. I kept running until I had a nice hot spot. Now, a hot spot isn't a big deal if you have the time to rest it and let the wound heal. Yet, I didn't know that. I probably should have known that, but didn't connect the dots. My pride kept me from admitting to my coach and running partner what happened during the run.

On the day of the marathon, I dressed in a cotton shirt and shorts and started at 5 a.m. By the time I crossed the finish line at 9 a.m., my thighs and nipples were bleeding from chafing, and I had an orange-sized blister on my foot oozing blood out of my shoe. My partner and coach said he'd never met anyone tougher or more stupid than me. Later, he retold this story at a Special Forces meeting, where he introduced me as the guest speaker. My stubbornness was not all bad; it would go on to get me through some demanding situations.

It was 1997; I was 28 years old. I had already buried two close friends, had lived in Ireland for two years, had joined the Army, and was married with three children. Life comes at you fast.

4

DE OPPRESSO LIBER (TO FREE THE OPPRESSED)

THE OFFICIAL MOTTO OF THE SPECIAL FORCES

"One hundred men will test today, but only three win the Green Beret"-
The Ballad of the Green Berets, Barry Saddler

I sought a better lifestyle for our young family, a better career, and, more importantly, fulfillment. I wanted to be on active duty, but it was 1997 and the middle of the Clinton drawdown era. The military was shrinking, not expanding. It did not seem likely that I would be able to get on active duty. My mentor was about to play a pivotal role in my life. I trusted him and felt like I could approach him with my dilemma.

One day, I asked him, "How do I get on active duty?"

He informed me that the only officers they were taken on active duty were in areas with critical shortages—doctors, lawyers, chaplains, and Green Berets. That made the road to active duty relatively simple: I would become a Green Beret. Next question: What is a Green Beret?

I knew Green Berets were commandos or something; my ROTC Department had posters of them hanging on the wall, but that was all I knew. This was before YouTube, social media, podcasts, and a proliferation of news about Special Operations, Special Forces, Green Berets, Rangers, or SEALs. Other than knowing that they were supposedly the best of the best or elite troops, I knew very little, to say the least.

In what has always been my way, I dove headfirst into the pool's deep end with minimal forethought. I put in a DA 4187 requesting a transfer from the 145th Field Artillery to the 19th Special Forces Group in November 1996. Both were units in the Utah National Guard, so it was easily facilitated. I was immediately assigned to The Special Warfare Training Branch (SWTB)—a 12-man Operational Detachment, Alpha Team or ODA. ODAs are the basic unit in all Special Forces and the smallest tactical operating unit in the Army.

SWTB's sole purpose was to sort out the "hopefuls from the hopeless." At the time, it felt like pure hazing. We were taught skills like map and compass reading, marksmanship, and planning with plenty of "smoke sessions." Smoke sessions were us getting our asses handed to us regularly from Friday afternoon to Sunday evening. Army Physical Fitness Test or APFTs, swim tests, land navigation with rucksacks, road marches, and obstacle courses—at the end of the weekend, there was zero left in the tank physically. While I was training, I set about learning everything there was to know about the Green Berets. I got my hands on several books like *From OSS to Green Berets, The Company They Keep, Lawrence of Arabia,* and *Five Years to Freedom.*

SWTB did have a legitimate role. Unlike the Federal component or active duty, the National Guard had limited resources and few slots to send soldiers to Special Forces Assessment and Selection (SFAS). Those passing SFAS were then invited to attend The Special Forces Qualification Course, or "the Q course" for short. SWTB's mission was to ensure they didn't waste time, money, or especially the few precious slots assigned to the National Guard. They had to ensure that the candidates they sent to SFAS would have the highest likelihood of passing.

The summer of 1997 was our Annual Training (AT), where Guard and Reserve soldiers trained for two full weeks. This one would be at Camp Dawson, WV, a small National Guard Camp nestled in a beautiful and inhospitable landscape of the Appalachian Mountains.

SWTB and their cadre put us through every test they could for two weeks. If we passed, they would allow us to go to SFAS. The heavy, thick vegetation and steep inclines were very different from the high mountain desert in which I was raised and had been most recently trained. But otherwise, I felt like SWTB and my personal training had prepared me well, and I successfully passed the internal SWTB mini selection; I would be given one of the precious National Guard slots for SFAS.

I wanted my attendance for SFAS to be just right: too early, and I wouldn't be fully healed from the mini selection I had just completed; too late, and I would lose the window of the physical hardness I had worked to achieve. Moreover, I wanted to avoid SFAS in the dead of winter and the heat of the North Carolina summer. Why add an additional level of misery and difficulty to what is known as the 24 most miserable days of your life?

I requested to go in the fall of 1997. SFAS was conducted at Camp Mackall, NC, a military installation adjacent to Fort Bragg, NC. At the time, Fort Bragg was known as "The Home of the Airborne" because of the 18th Airborne Corps and the 82nd Airborne Division, and it was also known as "The Home of Special Operations." At the time, Fort Bragg housed the 3rd Special Forces Group, the 7th Special Forces Group, US Army Special Forces Command (USASFC), US Army Special Operations Command (USASOC), Joint Special Operations Command (JSOC), and 1st SFO-D.

SFAS was everything the books, my SWTB cadre, and my mentor said it would be; it was no joke. The first phase was mental psych tests, personality tests, language aptitude tests, and IQ tests. Phase two involved individual events, including "The Nasty Nick" obstacle course. The Nasty Nick is named after COL "Nick" Rowe, who was held for five years as a POW in Vietnam and is the author of the book *Five Years to Freedom*. The course stretches over more than a mile of wooded terrain and includes more than 20 obstacles, involving underground tunnels and climbing rope ladders, just to name a few.

It was during this event that I failed to complete the Gut Buster obstacle. The Gut Buster was two logs running horizontally above the ground. The first log was about three feet off the ground, and the second was maybe eight feet off the ground. To successfully negotiate the obstacle, one had to jump up on the first log and then jump to the second log. Inevitably, slamming your gut into the second log, throwing your arms over the log, and then using upper body strength to pull yourself up and over the second log before hanging and dismounting. For whatever reason, I failed to complete it.

"Candidate report," barked the observing cadre with a clipboard in hand.

"Candidate 63 reporting, Sergeant," I replied.

"You failed to complete the obstacle, candidate; what do you wish to do at this time?" I was asked.

"Sergeant, I wish to renegotiate the obstacle," I stated firmly.

"Get back in line." He instructed as he made a note on his clipboard.

And I returned and got back in the line of candidates waiting for their turn to attempt the obstacle.

This process repeated for a total of three attempts, and each time, my anxiety of being dropped from SFAS increased.

On the third and final time, the cadre said, "Move to the next obstacle."

My stubborn, refuse-to-quit mentality forged in that sweaty gym, trying to succeed in wrestling—one of the few ways a kid could earn respect in that town—was paying off.

For runs of various lengths, the distances were always concealed so that candidates had to exercise a degree of strategy, choosing whether to run all out as if it were a 5K or to conserve energy for distances of 10K or greater. It also involved crossing Camp Mackall's many swamps, creeks, and acres of woods repeatedly, soaking our feet and boots, endlessly trudging.

The point was to cause our feet to soften and ultimately blister. No matter how tough you thought you were, if you blistered, every single step became a painful reminder and, more importantly, an invitation to quit. It also requires the ability to think clearly on your own, make tough decisions in ambiguous circumstances, and have a strong sense of self-motivation. Many candidates would "bust" their time deadlines, failing to make it back to their endpoint after successfully navigating to each of their designated waypoints. All too often, a candidate would sit down under a tree to change their socks and be tempted to close their eyes for five minutes. Five minutes would invariably turn into hours, and the cadre would have to search for them.

Phase three was the team events for those still in formation when phase two ended. Phase three was meant to assess how you performed as a leader and a follower on a team. When it was your time to lead, you led. When it was your time to follow, you followed. It was that simple, but it still caused problems for many. Some of the most physically fit and elite (usually guys from Ranger battalions) ran into a brick wall in this phase. Guys like this were headstrong and had elite pedigrees. When assigned the follower role for a particular task like moving an Army Jeep with only three wheels six kilometers, they couldn't take it if the appointed team leader from a logistics background, aspiring to become a Green Beret, struggled. They would want to take over, mistakenly thinking that the cadre who were watching (the cadre were ALWAYS watching) would see his natural-born leadership and somehow give him an attaboy. Or maybe reward him for getting the Jeep to the Rally Point (RP).

There was no necessity to move the Jeep. The Jeep was just a training aid or, rather, a leadership laboratory. In this specially crafted laboratory, on the sandy roads and snake-infested swamps of Camp Mackall, NC, the US Army Special Warfare Center and School (USASWCS) notoriously tight-lipped cadre carefully watched, listened, and wrote notes on each candidate and how well they performed their role. No leader in the Army is ever outside a "chain of command." There is ALWAYS, and I mean ALWAYS, someone above you and below you. You are both leader and follower 100% of the time, all day, every day.

Knowing how to lead when you are in charge and follow when you are not is key to a successful military career and is even more critical in Special Operations. Moving the Jeep just exposed us for what we were. It was like the wet boots the week prior, but tenfold. Already worn down, tired, fatigued with blistered feet and sore backs, bordering on illness, having dropped 10-20 lbs., nursing injuries, and hoping to hide them. You only slept 4 hours the night before and handed the Jeep mission at 0600 hours. You're one kilometer in when the wheels come off. And by "wheels come off," I mean BOTH literally and figuratively.

The pitiful contraption the team leader rigged to move the Jeep came undone. The Jeep is mired in sand; the team leader has lost the team's confidence, and half a dozen teammates offer advice. Another teammate is trying to take over to be helpful, and another half dozen or so, who are supposed to be pulling 360-degree security from an imaginary enemy who never attacks, are too preoccupied with the blisters on their feet, screaming at them to quit. It is here and now that you find out what type of Green Beret you really are. It was during these moments that we took note of each other for another critical aspect of SFAS: Peer Evaluations, or Peer Evals for short. Peer Evals were simple; we were asked to list each person on our team from best to worst and why for the top and bottom three. This was a simple but effective means for the cadre to be present by proxy and have 12 additional sets of eyes inside the team's inner workings. If someone thought he could hide, not pull his full weight, or was constantly whining or complaining, it was noticed. One bad peer eval was survivable, but if a pattern emerged, if the same guy was always in the bottom three and it was consistent across the team, the cadre would know where to focus. It was blood for sharks.

Our SFAS class started with nearly 350 candidates. On the morning of the final day, we were gathered into a formation under the bowery where we ate, adjacent to the dining facility or DFAC. There were maybe 80 of us, and despite this being the last day, not all 80 of us had passed or were going to attend the Q course. There is something known as a 21-day non-select. These are the most unfortunate bastards that there ever were. These are the guys who stuck in there, didn't quit, completed every task, and were on the bubble, as it were, right up until the final moments when the cadre made a final decision about their fate.

We were deeply invested and measuring ourselves against each other. Internally thinking or externally saying to small groups of peers things like, "There's no way that guy gets selected; he completely failed most of the obstacles on Nasty Nick." We tried to "war game" the outcomes and determine where we measured up. I guess it was a way to cope with our anxiety about possibly not being selected.

One of the cadres stood at the front of the formation and called out our numbers. Candidates are given numbers on day one, and the cadre would only address you by your number. It was a way to keep every candidate equal; there was no rank, no officers, no NCOs, just candidates.

"Roster number twenty-five over here," he shouted. "Roster number 132 over here," he continued.

And so on until they divided all 80 of us into one of three new, smaller formations. This only added to our anxiety as both types of candidates were assigned to each group—those we had convinced ourselves didn't belong and those that would definitely be selected. By the time your roster number was called, you weren't sure what formation you wanted to be in or what the hell was going on.

I have a clear and vivid memory of standing there waiting for my number to be called and being assigned to one of the three smaller groups, unsure of which group I wanted to be in. I remember telling myself: *This is it; I gave it my all, I did my best, and if my best wasn't good enough, I wasn't coming back to do SFAS again, fuck this shit.*

My roster number was called; I ran to my group and waited. The cadre finished assigning each of us to one of the three groups and then, with both arms outstretched and pointing to two of the three groups, shouted, "Congratulations, if you are in either of these two formations, you have been selected; you are dismissed for chow," then quickly pointing to the last group he shouted, "Quickly go to your huts, pack your shit and move to the 'shack of shame', you'll find boxes of MREs there for you." Just like that, SFAS was over, and I was selected.

SFAS doesn't train you to be a Green Beret. In truth, by the time you showed up to SFAS, you were either a Green Beret or not; maybe you just didn't know it yet. I subconsciously like to think I knew I was a Green Beret. Looking back, I can credit all the events in my life, like a 0-47 high school wrestling record and refusing to quit, running the St. George marathon with blood oozing from my shoe with every step, and carrying 100 lbs. of coal on my shoulders. I credit my "farm tough, never quit" mentality to my success at SFAS.

Later, those of us who were selected gathered in the large Auditorium and were addressed by the Senior NCO for SFAS. He said something that I'll never forget.

After a long briefing about our next steps, his final comment was "Selection never ends. Dismissed!!"

The implication here was that as a Green Beret, you are expected to adhere to a creed, to the ethos, to the Special Operations Forces (SOF) imperatives, to be the best, to maintain a standard, for the rest of your life. Otherwise, your rucksack would be placed in the hall. Placing a person's rucksack in the hall was the non-ceremonial way that an SF team would remove a soldier from the ODA, signaling that their performance had failed to meet the standard or that they had lost the faith and confidence of their teammates. If this happened to you, you would never recover. I later learned that of the men I finished with (65 of 350) and most of those I would serve with, our careers and paths would intersect dozens of times over the next 25 years, especially in combat that, unbeknownst to us, was looming on the horizon.

Following SFAS, I returned to my wife and children in Utah to begin to plan our next steps. I would need to plan when to attend the Q course. I wanted to attend class 2-98 in February 1998. There was only one small problem. I wasn't Airborne qualified yet, and many of the field training exercises were carried out using Airborne infiltration. I was told I would have to complete the three-week Basic Airborne school en route. Translation: I had to sell our home, move my family to Fayetteville, NC, and then fly to Fort Benning, Georgia, in January 1998, complete the three-week Airborne course, avoid injury, and return to Fort Bragg, NC, before early February. If anything went wrong and I couldn't attend the Q course on time, I would be ordered back to Utah to recover and have to pay for the move because I would be considered physically unfit for the orders to the Q course.

A special circumstance that only applied to someone in the National Guard was that if I were to get injured, the National Guard wouldn't pay my full-time salary while I healed up on active duty orders. They preferred a safer approach: send me to Airborne school on

Temporary Duty Orders (TDY) and return, then plan a date for me to attend the Q course.

All conventional wisdom and recommendations from everyone I knew were to slow down, go to Airborne school, return to Utah, then possibly enter Q course 3-98 or 4-98. Of course, I risked it all. I was on a mission and was confident in myself and my abilities. We went to Disneyland for Thanksgiving, returned home, put our house up for sale, enjoyed the most minimalistic Christmas holidays with all our belongings in boxes, and drove to NC shortly after the new year. A few days later, I flew to Fort Benning, Georgia, to attend Basic Airborne School.

I completed Basic Airborne School and returned to Fayetteville, NC, to start the Q course on time; my gamble had paid off, and I wouldn't have to return to Utah at my own expense. The Q course was challenging, exciting, demanding, and fun all at the same time. I was trained in small unit tactics, mission planning and leadership, and hand-to-hand combat. I was given weapons, explosives, communications, and first aid overviews. Additionally, we would be trained in unconventional warfare and guerrilla operations.

SFAS was designed to funnel the soldiers with the highest probability into the Q course. If you passed selection, you had the requisite physical and mental characteristics needed to pass the Q. Attrition in the Q course is still a real possibility but that is generally due to life events like having a change of heart when you learn that you will be deployed or away from family six to eight months a year, or a medical problem or violations with the Uniform Code of Military Justice (UMCJ) like a DUI.

I felt confident in my abilities, and as long as I worked hard, took care of myself physically, and was the "gray man[1]," I felt like I would earn

---

1. A gray man is a military concept of wanting to blend in, to not draw attention to myself.

the Green Beret. At the time, the "Pipeline" phases looked like this:

1. Small Unit Tactics (SUT).
2. Classroom for officers heavy on mission planning and leadership.
3. Robin Sage: the oldest and longest-running unconventional warfare simulation in the world.

Here marked the completion of the Q course, and we were officially awarded our Special Forces Tab and the Green Beret. Still, we had more events to finish successfully, or all of those achievements would be stripped from us:

1. Language school: I was assigned French.
2. Survival, Evasion, Resistance, and Escape school (SERE school): a training program for military personnel to learn how to survive, evade capture, and resist interrogation or a POW event, complete with physical beatings and light torture.

Most of us graduated from language and SERE and took command of an ODA by the Spring of 1999, or 2000 in my case. That meant that by 9/11, we were all the most seasoned and senior team leaders on ODAs sprinkled throughout Special Forces. Our class would include two team leaders in the initial invasion of Afghanistan (movies were made about this), more on the invasion of Iraq, several that went on to serve in Tier One units, several Colonels, one General Officer, one GS15 (the civilian equivalent of a General), three who would transfer to the CIA and several still carrying out classified operations to this day. The Global War on Terror (GWOT) had a propellant effect on everyone's military career. We all benefited greatly from being Green Berets and fighting a war suited for Special Operations for the remainder of our careers.

During language school, I was promoted to Captain and was now eligible to apply for active duty. I prepared a packet for active duty, complete with signatures from my CO and BN CDRs. My CO CDR felt like it was his duty to inform me that I "was shooting myself in the foot, and even if I were accepted onto active duty, that I'd likely never be promoted again."

Despite his doom and gloom, I moved forward and was asked to repeat many of the same events I had completed as part of SFAS, namely the APFT, 12-mile road march, psych test, interviews, and a board.

In the Army, a board is an interview in front of a group of interviewers where you must not only know all the answers but remain emotionally composed while the board members prod at your vulnerabilities the entire time. I passed and was given orders to attend the Infantry Officers Advanced Course (IOAC) with follow-on orders to my unit, 3rd Special Forces Group, Airborne (3rd SFG). I also asked for the opportunity to attend Ranger school in the process.

My orders were amended with Ranger school added between my SERE school graduation and the IOAC, with the caveat that I could not be recycled in Ranger school because the Army needed me to report to 3rd GP on time. Ranger school had remained a personal goal of mine, but as a "tabbed" Green Beret, it certainly wasn't a requirement. Like SFAS and the Q course, Ranger school was notorious for low graduation rates. Historically, the graduation rate has been less than 50%.

Ranger school at the time I went had four phases: 1) Benning Phase, 2) Darby Phase, 3) Mountains Phase, and 4) Florida Phase. I was part of class 8-99 and was sailing through Ranger school without too much difficulty until the Florida Phase.

The requirement is to pass a majority of your patrols in each phase. I was given one patrol in Florida and did not pass. Normally, a student is given more than one patrol, especially if they failed one and needed one or more to pass that phase. I was not given any more patrols, and subsequently knew that I was not going to pass.

While my classmates were getting ready to graduate and pin on the coveted RANGER tab, I was wrestling with the idea that I would be forced to start IOAC and possibly never get another opportunity to attend Ranger school again. I was called before a board that reviewed my performance in each phase, emphasizing the Florida Phase, and that I was being given the opportunity to "recycle." In other words, dropped from the course as a holdover, then added to class 9-99 following my original class. I am sure that 99% of those offered an opportunity to recycle accept it. After all, you can take courses on patrolling, recuperate, sleep, and eat while you wait for the class behind you; it is usually a no-brainer. But I stood at rigid attention and reported, "No, Sir, I can't."

The head of the board, the Florida Phase Commanding Officer (CO), a large, imposing man's jaw almost hit the desk.

"No," was not what he expected. He asked, "Why *can't* you?"

I reported that I was on orders to start IOAC in October and was told that I wasn't allowed to recycle, or I would miss my IOAC report date, and should I not pass Ranger school without recycling, then I would be forced to withdraw and make it to IOAC on time.

He promptly replied, "Wait outside."

I made an about-face, marched out of the room, and sat in the hall with no idea what was happening; it was black and white with no alternative courses of action. I sat there for maybe ten minutes when I was told to return. I followed the custom of knocking loudly and waiting to hear the "report" as the signal to enter the room. I entered the room, marched to the head of the board, and promptly saluted.

He returned the salute and said, "CPT Butler, Ranger class 9-99, will graduate on September 25th, and your orders are for IOAC class 4-99, which starts on Monday, September 27th, 1999. You will have 48 hours between classes. Now, I ask you again, do you want to recycle?"

Trying to hide my smile and maintain my military bearing, I replied, "Yes, I would like to recycle."

I was held over and put in class 9-99, where I easily passed the Florida phase on my second go. I graduated on a Saturday and reported to IOAC on Monday, still exhausted and starving.

Following IOAC, I returned to Fort Bragg and reported to my ODA, Charlie Company, 1st BN, 3rd SFG(A). It would be the best assignment I'd ever had. I would serve as team leader for a small team, trained to operate behind enemy lines and conduct the most dangerous missions with strategic implications. The team comprised some wild characters, but good men and great soldiers. Leading them was, and still is, the greatest honor of my military career.

My wife was jealous of my team, saying things like, "You love your team more than me," which was partially true, and I understood her perspective.

The teams form a bond from mutual agreement to fight and defend each other. There is a bond that can often be the strongest bond a soldier will ever experience.

When I would get ready to deploy, she would say, "You're already deployed," in the weeks leading up to actual departure.

She was insinuating that I was already on the other side of the world mentally. It is hard not to let your mind wander when performing real-world missions with life-and-death consequences in foreign countries.

In 2023, the officers of our class, class 2-98, gathered for the 25th anniversary of the Q course. As we gathered in one of our classmates' backyards in St Petersburg, Florida, we took a moment to go around the circle and share what we have been doing for the last 25 years. It was one of the most humbling experiences of my life, listening to the summaries of legendary men who all impacted global events for two and a half decades.

5

WAR

"I hate war as only a soldier who has lived it can, only as one who has seen its brutality, its futility, its stupidity." - Dwight D. Eisenhower

On Sept 11th, 2001, I was sitting in my Battalion Commander's office getting ready to brief him on a concept to conduct training in Jordan. We were both glued to Fox News as we watched the towers fall. He canceled our briefing and told me to go to my team room. It was a heavy day, as we all knew that this meant war.

The months between September and December were a flurry of training, planning classified briefings, and educating ourselves on Afghanistan. By December, all the pieces were coming together. We went with the entire Battalion (BN) to Fort Picket, VA, to conduct a full mission profile rehearsal where teams went through the entire sequence from notification, isolation, receipt of mission, planning, briefing to the BN CDR, mission approval, INFIL, mission, EXFIL, and debrief.

I was the senior team leader of what was the best team in the BN (so I was told). We were conducting a full mission profile to deploy to Afghanistan; I was ecstatic. I wasn't bloodthirsty, but I was eager to prove myself in battle—something I later would learn is an initiation process in most tribal and ancient cultures.

My former CO CDR, "Maj. G," had been moved up to BN staff and had a personality conflict with the BN logistician. Maj. G fired him a few months before the entire BN was deployed, and there was an enormous amount of work to be done. Maj. G called me into his office and explained he was removing me from my team and making me the BN logistician.

I vehemently objected. "I can't even spell logistician. I don't know anything about logistics. I am not trained in logistics," I explained.

"Look at your left shoulder; you'll figure it out," Maj. G stated as a wry way of saying, "You're a Green Beret," and the words "selection never ends" echoed in my memory.

I left my team, learning to be a logistician and deploy a BN with all of its kit with zero knowledge or experience.

I immediately hit the accelerator with my plans for departure. I had already begun looking into options for my post-ODA time. I had taken the APFT for the one located on Ft. Bragg, but chose to go in a different direction, believing that the other Unit I was considering was more diverse in its mission set and would be more enjoyable for me. By April of 2002, I was under selection for that Special Operations unit.

I passed and was ordered to attend the Operator Training Course (OTC). In many ways, I felt a deep sense of accomplishment and validation by dodging the S4 job, but on the other hand, I had abandoned my unit in its time of need, left it without a logistician, and had been removed from Special Forces altogether. This would be a decision I would regret repeatedly in my career.

While our family was getting ready to move, I heard a story of an NFL football player who had volunteered for the Rangers. His sacrifice and selflessness moved me. There is a tradition in the Army for those who go to Ranger School to carry a Ranger tab in their patrol cap. But it is better to have one given to you by someone already Ranger qualified. I knew Pat Tilman would have ample Ranger tabs given to him before he departed for Ranger School; after all, he was going to the Ranger Regiment. Despite that, I wrote a letter expressing my gratitude and respect for his decision, stuffed it in an envelope with one of my Ranger tabs, and sent it off to the Arizona Cardinals.

The OTC was even better training than I received in the "Q." More marksmanship, airborne operations, demolitions, communications, and first aid with a wide range of cross-training from former CIA operatives. I felt that, for the most part, I was holding my own. I wasn't the best in the class, but I wasn't the worst. It was much more challenging than the "Q," but I was surviving. Cadre counseled me on three occasions, and all three were typical of a unit like this.

I was simply told, "You are meeting all the performance standards; if you continue to meet the standards, you will graduate and become a unit member. What are your questions?"

I never had any questions. How could you have questions when the summary of three months of training was presented to you in 45 seconds by a stone-cold-faced cadre? We sat in the classroom on the last board, and names were called alphabetically. There were a couple of students who came before "Butler," so I had a good idea of when I would be called. They skipped me and went on to the "C" names.

I sat there for several hours. A couple more people were skipped. We all knew we were going to be dropped. Once they had reached the end of the list of classmates, they called me first since I had been the first skipped.

I was told, "CPT Butler, you failed to meet the standard. Thank you for all you've given to the course and the unit. Go home and wait for one of us to contact you to sort out your next assignment. Do you have any questions?"

Of course, I didn't have any questions. After all, I hadn't had a question for a year. I was beyond devastated.

I sat at home for six weeks without any answers. After six weeks, I decided to drive across town and go to the Army's personnel office. I shaved my non-regulation mustache, which I had grown during the course, found a set of BDUs I hadn't worn in a year, and drove to Alexandria, VA. I rode the elevator to the sixth floor of the SF branch offices.

I walked in and said, "Hi, my name is CPT Butler, and I need an assignment."

"Who are you?" they asked.

Eventually, I was ordered to join the Combined Joint Special Operations Task Force: Arabian Peninsula (CJSOTF-AP). This was shortly after the 5th GP invasion of Iraq, and I would be augmenting the CJSOTF staff. I was thrilled. I was finally getting into battle, finally on my way to Iraq.

We were attached to the CJSOTF from Aug 2003 to Feb 2004. My primary role was managing the Legion Security Force, or LSF. These were Kurdish fighters who had linked up with elements of the 10th Group in Kurdistan and had fought from the North all the way to Baghdad. They were now being used as Partner Force augmentees. We had occupied Saddam's Radwaniyah Palace Complex (RPC), about five miles southwest of the Baghdad International Airport.

I loved working with the Kurds or the "Muj," short for Mujahedeen (those who face death). I studied Unconventional Warfare (UW) and learned all I could while trying to get onto active duty. One of the heroes of UW was Lawrence of Arabia. If there was anything that I wanted to do in SF, it was UW. Working with the Kurds allowed me to live this out to some extent.

I moved my office from inside the palace with the rest of the staff to the shack beside the living quarters of the Kurds who were left on the perimeter of the RPC compound. I was there for seven months. It was a daily reminder about the fragility of life and how quickly life can change.

There is a first time for everything. The first time I heard the THHhhhWACK of a bullet caught me off guard. I was checking guard towers with an NCO where we had our Kurd allies when I heard it. It reminded me of the crack of a bullwhip we used with our cattle in my youth, but I didn't make the connection that it was splitting the air at supersonic speed.

"What was that?" I asked out loud.

"Bullet, sir." SFC Smith replied flatly. He had previous combat experience and was familiar with the unmistakable sound.

We crouched lower behind the sandbags. It was only harassment fire and not well-aimed, but it was enough to get my attention. It would be a constant in my life for the next seven months.

At night, we would take our position on top of the buildings within the compound. It reminded me of a song called *B.O.B. – Bombs Over Baghdad* by Outkast, but we called it "Bullets Over Baghdad." From our vantage point, we saw tracer fire[1] lighting the night sky. Various skirmishes and engagements all played out in front of our eyes. Sometimes, they would be directed at us; most of the time, not. There was no real way of determining how lethal it was. To me, it felt random; the whole war felt random.

Another example of the randomness of our situation was the nightly

---

1. Tracers are projectiles coated in a special paint, that when fired, the paint ignites and burns brightly, making the projectile's trajectory visible.

mortars and rockets aimed at our compound. Considering that we slept in tents, they posed a real threat. In my mind, I pictured Monty Hall with a thin microphone standing on a stage, sweeping his hand toward our Fire Operating Base (FOB) and, in game show voice, saying, "Welcome to tonight's episode of *Where Will the Rocket Land*." Fortunately, no one was injured.

In its infinite wisdom, the Army had multi-million-dollar machinery called Counter Fire radar. It was supposed to intercept the trajectory of the rockets and, based on calculus and physics, determine the Point of Origin (POO) of the rocket. That information was fed back to an Artillery unit, which was then assigned to fire at the POO. There were two problems with this concept. The first problem is that the enemy figured out quickly that our Rules of Engagement (ROE) didn't allow us to fire within the vicinity of NO-GO places like mosques, schools, and hospitals, so they would fire from there. The second problem is that they became inventive, creating rocket-firing platforms mounted in Toyota pickups or hand-held mortar tubes they could easily disguise and fire from motorcycles. They could fire and move quicker than our expensive machine infrastructure could return fire.

We had to innovate. SFC Carlson, who had been a sniper instructor at the infamous Range 47 on Ft. Bragg, volunteered to handle it. As a team, we sat down and plotted the areas that held the most POO sites. We analyzed the most frequently used POOs and viable avenues of approach for SFC Carlson. We then would load up in our tactical vehicles, forming a convoy, and use a "humanitarian" or a "village elder visit" to drive out within range of these POO sites. When we would get to an unobservable spot in the road, SFC Carlson would tuck and roll out of the Humvee into a ditch, where he would stay covered and concealed for up to three days, observing and firing when the opportunity presented.

During those nights, we would remain ready to roll out and recover SFC Carlson at any moment, should he call for backup. Shortly after we employed this tactic, the rocket fire diminished, moved off to greater distances, and its accuracy was significantly reduced.

When I returned to Virginia, the COL sat me down and said, "CPT Butler, I am going to make you the morning Battle CPT."

The Battle CPT was a single position present during each of the three daily shifts: one from 0400 to 1200, another from 1200 to 2000, and a final one from 2000 to 0400. The downside of the 0400 shift was obviously the hours, but the upside was that I would be home when my kids got home from school and could spend more quality time with my family. The role was fairly simple—track all the FWD deployed units, ensure the morning slides were arranged into a daily briefing presented to the CDR and his primary staff at 0800, monitor the entire daily operations of the BDE until 1200, then conduct a "hand-over" with my follow-on counterpart who would essentially repeat the entire process, less the daily briefing.

I had been in the role for under six months when I processed an email with a Request for Support (RFS), a planner with a Ranger tab to support the 75th Ranger Regiment on an upcoming deployment in 2005! The unit has only two officers with Ranger tabs: the BDE CDR and me. A few weeks later, the BDE CDR called me into his office and broke the news.

"CPT Butler, I know that I promised you a year of downtime to keep you with your family and give you some much-needed time to recover from the speed you have been running, but we have this RFS for a planner with a Ranger tab."

I cut him off there and replied, "Yes, Sir. I know, Sir."

By March 2005, I would find myself on my second combat deployment since 2003 and attached to the 75th Ranger Regiment S3. They performed airfield seizures and raids better than anyone in the world. If the POTUS wanted to breach any country, seize an airfield, create a toehold, and begin flooding divisions of Infantry and support troops into said country, the Ranger Regiment was who the POTUS would call.

At that time, Special Forces had five distinct missions: 1) UW capable of living behind enemy lines operating in the native language, 2) Direct Action (DA) tactical actions on an objective, same as the Rangers, 3) Special reconnaissance, 4) Foreign Internal Defense, and Counter Terrorism (CT). The Rangers were the 10lb. sledgehammer and Special Forces was the surgeon's scalpel. Also, the average age of a Ranger was 21, while the average age of a Green Beret was 30. The average rank of a Ranger was E4, while the average rank of a Green Beret was E7.

The rivalry fluctuated between healthy and dysfunctional. Contributing to the tension, many Rangers sought more training, better rank advancement, and a higher degree of variety, naturally progressing to Special Forces. Conversely, Special Forces, by design, attract the best and brightest. The standards for testing, IQ, personality, physical performance, and leadership are the highest in the Army. When the Ranger Regiment, a relatively small unit, loses some of its best and brightest to SF, tension is created.

With this tension, I walked into the Tactical Operations Center (TOC) in Afghanistan and met Maj. Renyolds. To his credit, Maj. Renyolds was all business and didn't have time for bullshit. He just wanted my best performance and was happy to have his RFS filled. Not everyone else was as welcoming. I ate alone, worked out alone, and spent my time off alone; there wasn't a Ranger around willing to befriend me. I had been advised to remove my SF Tab from my shoulder before deploying, but I refused.

In a strange turn of events, I was there for the one-year anniversary and the ceremony of Pat Tillman's death. Although I never served with or knew him, my mind traveled back to 2002 when I sent him my Ranger tab. I couldn't help but think about the strange coincidences of life.

When I returned from Afghanistan in 2005, I went home to Utah for a visit by myself. Looking back, I can see how I was dealing, in my own way, with my demons. During that visit, I stayed up in the mountains, where our family kept a couple of camp trailers. My uncle was a Vietnam vet and met me there. For almost a week, we sat beside the fire and didn't say a word—just stared into the fire. Civilians wouldn't understand that, but vets do. It isn't important to talk about it so much as it is just to be around people who understand the desire not to talk about it. Unknowingly, I was using nature to process.

In September 2005, I was promoted to the rank of Major and later given orders to attend the Command and General Staff College, or CGSC, from 2007 to 2008. This was the first professional development in a Field Grade officer's career, marking a significant shift in career trajectory. Following CGSC, I wanted to return to 3rd GP as soon as possible, but my chances weren't good; I had been out of SF for six years.

On the morning of graduation from CGSC, I was walking into the building when one of my SOF instructors saw me and yelled, "Hey, Butler, come here."

"Yes, Sir?" I said.

He asked, "What is your assignment following graduation?"

I told him that I had been assigned as a desk officer at USASOC.

He simply said, "Do you want a company command?"

"YES!" I said.

He pulled out a piece of paper, wrote a number on it, and said, "Call this number ASAP and tell them that I told you to call them; tell them you want the company command."

"Yes, Sir," I replied.

3rd SFG (A) Group Support Company, or GSC, wasn't a "Line Company" made up of ODAs, but it was back in Group, and I was forever grateful. GSC opened my eyes to the role of support soldiers in the Special Forces machine. GSC comprised engineers, medics, signal, military intelligence, a drone platoon, and a chemical detachment. It was also an honor to serve with the often overlooked and underappreciated support soldiers of GSC who deployed every six months—more than anyone else in Group.

I was reunited with my former senior weapons MSG Rodney; he would be my first sergeant (1SG). I was grateful to serve with him again and knew I could trust him implicitly. He was a quality soldier and had been in Group since I left. We would deploy the company twice during my two-year command.

As a family, we moved four times in nine years, and I deployed to Iraq or Afghanistan four times. I was not accounting for the strain that I put on my wife or children.

Following my command of GSC in the summer of 2010, it was time for a new assignment. I sought out an assignment in the Joint Special Operations Command or JSOC.

Although I was applying for a staff position within the J7, there was still a selection process, and if successful, I would be assigned to a three-person team as an exercise planner. In this role, our team would spend almost a year planning a single exercise for subordinate Tier One units.

While serving in JSOC, I was fortunate to serve under the commanders Adm. McRaven and Gen. Votel. I was always spoiled with great leaders.

In 2011, I would go through my first divorce. After 20 years of marriage, we became roommates, and worse yet, I was not faithful. As a Mormon, this meant only one thing—excommunication.

Experiencing excommunication from a church like the Mormon church can have profound psychological and emotional consequences. The trauma includes, but is not limited to, 1) loss of identity, 2) loss of community, 3) public judgment, 4) internalized shame, 5) depression and anxiety, and 6) grief and mourning, to name a few. It is a barbaric, outdated, and shame-based practice rarely practiced by any organization in contemporary society.

This would result in our divorce and couch surfing with friends from work. I only had a couple of bags of clothing, with everything else in storage. I bounced from couch to couch for six months. We negotiated our divorce at a coffee shop, took it to our respective lawyers for their opinions, and then backdated the date of separation to October 2010, allowing for the divorce to be final in October 2011.

In 2012, I was about to make the classic post-divorce mistake, the mistake of a rebound marriage. Just over a year after I separated from my wife of 19 years, I married Jessica. In less than 90 days, there was already some serious friction. Unlike my first marriage, I suggested marriage counseling. I was trying to be a whole new me. Old Matt saw therapy as a sign of weakness; new Matt was trying to be a better husband. She had stopped responding to "I love you." I was optimistic that counseling would help lay the foundation for a happy marriage. On day 91 of our marriage, we walked into the marriage counselor's office and sat down.

The counselor asked, "Matt, what do you want to achieve from counseling?"

I hoped to learn how to better communicate with Jessica to build a solid foundation for a happy, healthy relationship and have a long and beautiful life together.

The therapist responded, "Wow, that is the best answer I have heard in 20 years!"

She turned to Jessica, "Jessica, same question."

Jessica didn't respond immediately.

After several attempts from the therapist, Jessica responded with, "I want a divorce."

I moved out the following week. The week after, I volunteered for Afghanistan and my fifth deployment. War was easy, but life was hard. When I was deployed, all I had to do was wake up, work out, do my job, sleep, and repeat. It became an escape for me.

I walked into the lawyer's office, who had done my first divorce. He was surprised to see me so soon and joked about getting the "next one for free."

I handed him a prenup, wrote a check, and said, "Email me when it's final; I am going to Afghanistan." I never saw or spoke to my ex again.

Soon, I would be on my fifth deployment and fourth to Afghanistan. Interestingly, despite this being the most preeminent and elite unit in all the DoD, they did not have unilateral authority to conduct missions. In our building, a JOC of combined US personnel (Navy, Army, Marine, Air Force) was on the top floor, while an Afghan-maned JOC was operating on the ground floor. I led a team as the US liaison between the two JOCs. When intelligence reached a critical threshold, the mission was passed to the Unit carrying out the mission, who would pass it back to the Task Force for approval.

If the Task Force approved it, a redacted version was passed downstairs for Afghan approval. There was one representative from each of the four main elements in the Afghan Defensive Organization—The Ministry of Defense (MoD), the Afghan National Army (ANA), the National Defense Service (NDS), and the Afghan National Police (ANP). Each member of the Afghan JOC was allowed to review the mission and either approve or disapprove the mission. One single "disapprove," and the mission was scratched.

On more than one occasion, I had to go upstairs and explain why one of the four Afghans had disapproved. Those were tense meetings filled with Senior Field Grade Officers, General Officers, and representatives from our most elite units who had men sitting in HELOs, waiting for a call, allowing them to make their "hit time."

After I returned, I was made the Deputy J7. Instead of being on a team of planners, I now oversaw four teams of planners. In Sept of 2012, I was with one of the Deputy CDRs conducting a level two exercise in the UAE. We ran a 24-hour operation in 12-hour shifts. As the senior exercise planner on-site, I took the night shift. On the night of Sept 11th, I received a phone call with an accompanying email that had the following header:

"REAL WORD//REAL WORLD//REAL WORLD// Not part of the exercise."

Terrorists had attacked our Embassy in Benghazi, Libya.

I yelled out over the Joint Operations Center (JOC) for everyone to "shut up" with an intensity often saved for combat TOCs. JOCs are kept in Special Compartmentalized Information Facilities SCIFs for short. Exercises were far more lackadaisical. I am sure everyone wondered why LTC Butler was so intense for a night shift during an exercise. I had my ear pressed down, holding the receiver to my shoulder while I scribbled notes in my green standard-issue notebook.

"Uh huh . . . Okay . . . Yes, Sir, uh huh, Roger," was all I could respond with as I took notes.

When you need to get everyone's attention to relay vital information, the standard protocol in a JOC is to yell, "Attention in the JOC!" At that point, everyone stops what they are doing and focuses on the person who called it.

"Attention in the JOC," I shouted.

Everyone turned, facing me, and froze, stopping whatever they were doing.

"We have a real-world crisis. I just forwarded it to each of your inboxes. I'll need you all to read it and start war gaming scenarios, and I will wake up the General. What are your questions?"

There was silence; I assume everyone was more interested in reading the email.

I made my way to the Billeting, where the General had his quarters. In war and exercises meant to mimic war, there are "wake-up criteria." It is intended to ensure that leaders aren't woken up during 24-hour OPs and prolonged operations for nonsensical reasons. This was definitely within the "wake criteria." The General would know it would be important if I knocked on his door.

Knocking loudly on the door, I yelled, "Sir, it is LTC Butler. We have a real-world incident, and I need you in the JOC ASAP."

Blurry-eyed, the General cracked the door and asked, "What is going on?"

"We need to discuss it in the SCIF," I replied.

"I'll be right there." He said, shutting the door to get dressed.

The General arrived at the SCIF and immediately called JSOC. After a short conversation between us and a phone call back to Ft. Bragg, the General called "Attention in the JOC." The exercise was canceled, and we shifted into Crisis Response immediately. Our staff shifted seamlessly into crisis planning, developing a course of action that would go to the General on-site for approval. For the next two days, we watched in horror as events unfolded. We had UAV feeds, emails, and phone calls. We pushed as hard as we could for any of our COAs to be approved, only to have them all denied.

All in all, during my time in JSOC, I was able to be a witness or part of Osama Bin Laden's killing, the Benghazi debacle, and the Capt. Phillips rescue. I like to phrase it this way: "I may not have been at the center of history, but I was on the sidelines with the best seats in the house."

6

WALLY

"Owning our story can be hard, but not nearly as difficult as spending our lives running from it. Embracing our vulnerabilities is risky but not nearly as dangerous as giving up on love and belonging and joy— the experiences that make us the most vulnerable. Only when we are brave enough to explore the darkness will we discover the infinite power of our light. - Brené Brown

"I had to shoot him." His words came out slowly, deliberately— forced even. They filled the room with palpable tension.

I sat completely still, not uttering a word, allowing the void to hang between us, optimistic that the obvious silence would encourage Wally to say more.

"I had no choice; he is the one who still haunts me. I see his face when I sleep."

It was 2012, and I was struggling badly. My mental and physical health were in shambles; hell, my whole life was in a shambles. I had five deployments, was wrestling with an alcohol problem, and was fresh off my second divorce in two years. Two days after being told of the pending divorce, I ran into my good friend Scott, with whom I served in the 3rd GP Special Forces. Our kids grew up together, and we shared the Mormon faith. Scott took me to Krispy Kreme, and I spoke to him for an hour about how my life was unraveling.

I concluded my explanation only for him to say, "This is awesome!"

"Excuse me?" I said.

Scott answered, "Matt, I just submitted my retirement papers and am moving to Michigan in a few months. I have been concerned about who would look after Dad. This is perfect. You can move in with Dad."

"Dad" was Scott's dad, Wally. I knew Wally personally. He, too, practiced the same faith, and I had spoken to him many times on Sundays. I knew that Wally had also served in the Green Berets, and his son Scott, my friend, had followed in his footsteps, but that was the limit of my knowledge.

Without missing a beat, I replied, "No. No way, Scott."

I could barely look after myself in those days. In the back of my mind, all I could think was: *yeah, this is just what you need, to be a nursemaid to an 82-year-old geriatric.*

Scott said, "Just think about it, Matt."

I told him that I would, but in truth, I had no intention of considering it at all.

I had a short deadline to be out of the house with very few options.

"Scott," I said.

"What's up, brother?" He replied.

"I have been thinking about it, and I think I might be interested in living with your dad," I replied.

Scott was ecstatic. Mind you, I made sure to leave several escape clauses. I put everything into storage, made sure not to sign a lease, refused to pay a deposit, and told Scott, "If there is one single issue, I am out of there, no questions asked."

Scott replied wryly and said, "Okay, brother, no problem!" Obviously, he knew way more about how this would go than I did.

Wally would be up at 0400, and I would be up at 0500. I would make oatmeal for him and feed the dog before leaving for PT and work. At the end of the day, I'd return home, make dinner, do the dishes, and tidy up. On weekends, I would mow the lawn, do the shopping, and take care of the trash and other chores. We ate dinner in the living room in the EZ Boy recliners while we watched *Bones* reruns. We called the house the "team room." To my surprise, we were roommates and on the fast track to becoming great friends.

As the trust began to build, Wally began to tell me his life story. Wally had joined the Idaho National Guard at 15; there were different rules back then. One year, while at Annual Training (AT), his family moved from Idaho to California. When Wally returned from AT, he found a note pinned to the barn. His family had moved to California.

It didn't take long for Wally to grow tired of California and seek out an Army recruiter. Before long, Wally was sworn into active duty and shipped to Japan as part of the 1st Cavalry, the 24th Infantry Division, in late 1949. This would be the first of several fateful events in Wally's legacy. A few months later, Wally found himself in Korea, ten days after hostilities broke out in June 1950.

By April 1951, Wally would be wounded and receive the 1st of his four Purple Hearts. By Jan 1953, he had reenlisted and returned to Korea. As time wore on, I came to know what a legend Wally was in a community filled with legends. Wally fought in Korea and Vietnam twice each. His decorations included the Purple Heart with 3 Oak Leaf Clusters and multiple Valor awards. He knew Roger Donlan, a

Medal of Honor Recipient, Barry Sadler, a Green Beret singer-songwriter of The Ballad of the Green Berets, and Nick Rowe, the POW and namesake of Nasty Nick.

Most nights, we would sit, basking in the glow of the big screen TV, mostly in silence, occasionally talking about our favorite topics—golf, dogs, and guns. One day, our conversation took a different turn. After covering the usual topics, Wally said something that caught me off guard: without warning or segue.

"There was this one battle," he began. "It was early in my first tour to Korea. We had just fought a several-day-long battle against the Chinese with heavy casualties."

I quickly assessed what was playing out and that Wally was about to open up to me.

He continued, "We were searching for our dead and wounded, going from foxhole to foxhole, clearing them one by one."

I quietly acknowledged with an "uh huh," careful to say just enough to encourage him but not to say too much as to interrupt.

He paused, took a deep breath, and said, "I was by myself when I popped up over this one foxhole."

My heart rate was climbing as the gravity of this moment settled in.

Wally's tone dropped as he continued, "I peered over the edge, and I saw him staring back at me with terror in his eyes." He paused again, choosing his words slowly. "He was wounded, a young Chinese soldier; he couldn't have been more than 16, just a kid."

My throat tightened as I swallowed my emotions in complete silence. Wally paused for another long period. He was beginning to choke up.

"His guts were completely hanging out, and he was covered in dirt and blood. He had gangrene."

Like me, Wally grew up on a farm, and every farm boy knows what gangrene is; he knows the smell. I knew where this story was going as the weight settled onto my chest.

Another long silence, and then he said, "I had to shoot him. I had no choice. He is the one who still haunts me; I see his face when I sleep."

I froze and listened as Wally concluded his initial effort to open up to me with, "I have never told anyone that story, not even Scott, and I am not sure why I can't."

And just like that, the next episode of *Bones* was queued up, and nothing more was said that night.

After that, our conversations slowly shifted from dogs and golf to more weighty matters. We talked about war, life, love, and death late into the evening. We began sharing things that neither of us had said, and no other human had ever heard until then.

Wally, the consummate leader that he was, set the example for me. He knew there were things inside of me that needed to come out, that needed the light of day. He knew that when we give utterance to those things that most of us want to keep buried, they would no longer hold power over us. On that day, Wally was the "breach man" (a person on an operation whose role was to breach a door to allow the team to flood into a building). He created a safe space framed in trust and respect and free of judgment. I would cautiously begin to dip my toe into that space.

Wally would remain a key figure in my life for another twelve years. He was a father figure to me. I do not know when I realized this, but there came a point when I knew who the real nursemaid was and who needed nursing. Wally carefully, slowly, methodically nursed me back to health, both physically and emotionally, over those eighteen months.

Soon, I was in Afghanistan, my sixth and final deployment, with follow-on orders to SOCOM in Tampa.

At this point, I had been seeing a therapist for two years. I still wouldn't agree that I had PTSD, but obviously, something was wrong. Mistakenly, I thought it was the only option and put on antidepressants. At one point, my therapist suggested that there was more to my PTSD than combat.

She strongly urged me to find my birth parents and reconcile my abandonment issues. I thought it was unlikely that I would find them, but I registered my name and birth information on multiple adoption sites that seek to reconnect families. I wasn't very successful on my own. After a few months, I gave up and hired a professional. Within a week, they had my birth mother's name, address, and telephone number. I didn't wait to contact her, but I called the number, and it went to voicemail.

"Hi, I think you have been looking for me. My name is Matthew Butler, and I have some information that says I am your son. Anyways, my number is XXX-XXX-XXXX, please call me back."

In less than an hour, I spoke with Jannice. She had recently retired and was about to move from Utah to Washington. My tradition was to always go home to Utah and see my family before a deployment, as there was always the very real possibility that it might be the last time I saw them. We quickly planned to meet while I was in Utah, requiring her to postpone her move by several weeks. We met in Park City for coffee and kept the meeting very informal. Just a "get to know you" and putting a face to the name type of meeting.

7
ALL GOOD THINGS MUST COME TO AN END

*"Every new beginning comes from some other beginning's end." -
Seneca*

My five previous combat rotations lasted four to eight months. I was sitting at five deployments with 28 months of total deployed time when I volunteered for this sixth and final deployment. I was running away from my divorce and Fayetteville, North Carolina, which was full of bad memories for me by now.

The US Special Operations Command (USSOCOM), located at McDill AFB in Tampa, Florida, was a giant top-heavy command that oversaw all Special Operations for all four services. It was a behemoth of an organization. Going there as an LTC would make me very low on anyone's radar. As I found out, there was a glut of Field Grade officers who were assigned there and weren't deploying. How they managed to decline orders to combat was beyond my comprehension, but it was a fact. I learned that if I were willing to volunteer for a combat tour, I could choose a three-year assignment within SOCOM. I did. I volunteered for a one-year assignment with NATO Special Operations Command

Afghanistan, or NSOC-A. I would deploy from June 2012 to June 2013.

Gen. Austin Miller, a legend in the SOF community, commanded NSOC-A, and I had several close friends going on this deployment. I was a member of the Joint NATO Special Operations Unit, meaning there would be NATO allies, Navy SOF, Marines, and Air Force SOF. It would last twelve months—something I was not prepared for.

One of my first challenges was detoxing. In past deployments, I had been far enough away from the "flagpole" that I could drink and hide it well. In 2001, as part of the original GWOT, the DoD had enacted something called "General Order Number One" or (GO-1). GO-1 was a directive that outlined prohibited activities for US military personnel. Prohibitions typically included alcohol consumption, drug use, fraternization, possession of pornography, gambling activities, and personal pets. On the smaller Forward Operations Bases (FOBs), getting away with breaking GO-1 was easier. Drinking and porn were the most broken rules. NSOC-A would be different. We would be on a very small compound with a high density of senior NCOs and Officers, including Generals. We would be living in very close proximity. I took my last sip of alcohol before that deployment at dinner the night before I shipped out. When I arrived, I promptly went into a full-blown detox. I wasn't performing well at my job, struggling to sleep and eat, and battling cold sweats and shakes. This continued for over a week, but I passed it off as a "bug" and "jet lag." I hadn't recognized the severity of my drinking until this episode.

While there, I would get an email and a letter from my lawyer with sticky tabs indicating where to sign. And just like that, I was divorced for a second time.

In April 2012, I met a woman named Mallory online before deploying. Returning to my Mormon roots made logical sense, and I joined a Mormon dating app. Mallory and I would text and talk on the phone occasionally as I was getting ready to deploy. Our relationship became more intense when I was officially divorced, and I began to plan my mid-tour leave. The first week, I would spend with my family, including my kids in Utah, and the second week, I would fly to Delaware and spend time with Mallory. It would be the first time meeting her and her four children here.

Throughout my one-year deployment, Mallory and I corresponded through emails, letters, and phone calls. After the mid-tour leave, I felt even more encouraged. After all, she was Mormon; we connected, and everything seemed to click. However, I was still stinging from my last divorce and had internalized the lessons of a quick courtship and decided to take it slow with Mallory.

When I returned in 2013, I realized that if I were serious about dating Mallory, living in Florida and her in Delaware wouldn't be realistic. So, I began to look for a way out of my three-year commitment to SOCOM. I had only been there a few months, and I was already unhappy with the top-heavy nature of the unit, as I said: as an LTC, I was nobody. I knew of and had friends who served in the Asymmetric Warfare Group (AWG) unit at Ft. Meade, Maryland. It was still a 90-minute drive between Ft. Meade and her home in Delaware, but that was better than flying between Florida and Maryland once or twice a month.

AWG was the brainchild of Gen. Votel prior to his Commanding JSOC. The Pentagon had tasked him to figure out how to stay ahead of the enemy when it came to improvised explosives on the battlefield. Usually, the Army fields new equipment through a structured process called the Army Modernization Strategy. It involves several key steps: 1) research and development, 2) testing and evaluation, 3) approval and funding, 4) production, 5) fielding via awarding a contract, and 7) training. This process can take years. The enemy was shifting tactics monthly, if not weekly. It was Gen. Votel's job to get ahead of them.

AWG was given unprecedented latitude to examine the enemy's tactics and move through the same five steps in a matter of months. Unique to this process was the utilization of off-the-shelf technology and equipment, and teams of master trainers were sent down range and embedded with the units facing the new tactics to field and train on the new counter-tactics. The unit was a strange mix of about 50% senior officers, senior NCOs, and 50% contractors, many of whom came from 1st SFOD-D, Special Forces, and the Rangers. While not "classified", the unit consisted of many who had served in those assignments and, therefore, was built around a similar framework. In other words, before being assigned to AWG, I'd have to attend and pass another selection process, making this the fifth and final time in my career.

I would have assumed I had been through enough selections by now. My time in the unit was mostly great; I was assigned to the operations or "3-Shop," where I assisted in orchestrating the major muscle movements for the entire organization for the 1st year. I also got to deploy with a team to Lebanon for an evaluation. This would be my last overseas trip in uniform, and I absolutely loved Lebanon.

I was moved to the leadership development team during my second year at AWG. We were responsible for training the leaders at each of the Centers of Excellence (COEs). These centers are where soldiers in similar career fields are grouped and trained in their initial training before being sent to their units. They then return for additional leadership training throughout their careers. We would be training the Leadership Cadre at each of the COEs. In addition to developing new tactics, AWG, with its heavy population of SOF forces, also saw an opportunity to export the leadership style and lessons that had been sharpened by heavy combat to the rest of the Army.

I intended to retire from AWG in 2018, but the Army had different plans. The reality of my decision to go to the classified unit immediately after being removed from my ODA prior to their deployment was catching up to me. I had been passed over for Battalion (BN) Command. I was told I would never get a BN CMD because I hadn't served in the critical roles of Executive Officer (XO) or BN S3. The Army follows an up-or-out policy; if you aren't getting promoted, you are getting put out to pasture. Without a BN CMD, I'd never make Colonel (O6); I would not be promoted, so the writing was on the wall. I was, however, a senior Lt. Col., which made me fodder for the Special Forces Branch. Guys like me, who are non-promotable and close to retirement, could be assigned anywhere those groomed for higher positions didn't want to take. That was the situation in April 2106 when my phone rang.

"Hello, this is LTC Butler," I answered.

"Matt, it's Martin." I heard on the other end. Martin and I served as team leaders and CO CDRs in the 3rd GP, and he was the SF branch chief responsible for all SF assignments.

"What's up, Martin?" I asked, not anticipating that it was a call about my future.

"I need to reassign you; you'll be coming out on orders for a move this summer." He informed me.

"No, thanks, Martin, I am good; I plan to retire out of here in two years," I replied, unaware of how little leverage I had.

"Nope, you'll hit three years at AWG soon, so you're in the window for reassignment." He countered.

"I really don't want to move, Martin," I replied adamantly.

"I don't have any choice." He was more adamant. I knew he was right.

I asked, "What are my choices?" assuming I'd be given a list of assignments. Assuming I might be lucky enough to get a juicy assignment to Europe or something.

"The Pentagon." He replied.

"That's it?" I replied in shock.

"That's it." He stated firmly.

"You'll have my retirement paperwork on Friday," I concluded.

Taking an assignment at the Pentagon would mean waking up at 4 a.m., commuting for 2 hours, working from 6 a.m. to 7 p.m., and commuting back another 2 hours. My life would have effectively ceased to exist outside of work. I wasn't willing to accept this level of drudgery this late in the game, after 26 years of service, 40 months in combat, and countless days away from home and family.

Unceremoniously, without much leverage, it was as if Martin had taken an old dog out behind the barn and shot it in the head; my career was over.

A soldier can submit retirement paperwork for up to a year out. I submitted mine for the full year. I'd spend the next year balancing my work responsibilities while completing the Army's retirement process and planning my next move. Things were about to get worse.

The thing about mental illness is that it attracts mental illness, and they often don't play well with others. Also true about mental illness is that those with it can often mask it in short bursts. It is my sincere belief that both Mallory and I were wrestling with PTSD. Although Mallory and I had known each other for two years before getting married in 2015, things were really bad by the summer of 2016. It was so bad that we got into a fight that resulted in mutual "no-contact orders" and forced separation.

She had drained both bank accounts and terminated our credit cards, so I was left with $6 in my wallet. I couldn't afford a place to sleep, and the gas I had in my tank was gonna have to last until I could make alternative plans. I went to the only other place I could think of: my office. At the time, our unit was housed in what used to be a Brig or military prison, complete with gallows. It was a pretty sketchy place, but in haste to stand AWG up as a unit, the decision was made to take an existing building out of mothballs rather than spend the time building one. This Brig had a vacant basement; we mainly stored office furniture down there. I rearranged the furniture until it resembled a bedroom, found a cot, and used my issued sleeping bag.

Two weeks later, I moved in with my friend and co-worker John. John had recently gotten sober and was now running a sober house. Six men were sharing a six-bedroom home in the Baltimore suburbs. The requirements for living there were maintaining sobriety, attending AA meetings at least weekly, and having a sponsor. I hated AA meetings; my life was circling the drain for the third time. I was filled with rage, and it showed. When I did share, it was from a place of controlled rage, allowing the others in the circle to glimpse the rage boiling within, but at least I had a roof over my head, some support, and regularity.

The timing of the no-contact orders couldn't have been worse. As a soldier with dependents, a wife, and four stepchildren, there were administrative decisions concerning them that required us to communicate. At times, we used the courts as intermediaries. Jay likes to tell the story of how severe my crazy got.

On one occasion, with the court's help, I tried to arrange a time to go to the house and retrieve my belongings. I had several communications with law enforcement, and the courts confirmed the appointment. I had arranged a team of co-workers and neighbors to help with the move; everything was arranged. At about 10 p.m. on the day prior, I received a message from Mallory through a third party telling me, "Tomorrow wasn't going to work for her."

That evening, a massive storm, full of thunder and lightning, raged outside. It matched the storm raging inside. I wasn't able to keep a lid on my rage any longer. I went outside, only wearing the thin sweatpants I was wearing before bed, and spotted the exact tool I needed in the moment.

There, at the base of a row of trees, was a waiting axe. I picked up that axe with all the muscle memory I had formed in my childhood and started swinging with every ounce of rage, giving it an outlet. I chopped and screamed at God. I dared him to kill me.

"Fuck you. Fuck you. Fuck you. Fuck this life. I fucking dare you to strike me dead. Come on, I dare you! What kind of God can't direct a little lightning my way?"

I continued to yell with the steel axe head pointing up at the sky, hoping it would attract the electricity that filled the sky.

I continued to chop in full downpour. By now, I was completely naked, swinging the axe at every log I could find, rain pouring down, thunder and lightning all around me. I finally ran out of steam and fell to my knees in the mud and rain and sobbed. When the crying ended, I stood up and turned around to discover that my five roommates had been watching in a capacity to ensure my safety. The horror on their faces gave me an idea of just how scary I must have been at that moment. One of them later told me that he was worried about me, but so frightened that he wouldn't have known how to intervene if he had to.

The no-contact order ended in January 2017, and I started out-processing and terminal leave in March 2017. The wise decision would have been to file for divorce and run for the hills. But the shame of another divorce overcame me, and my pride wouldn't allow me to do the rational thing. When Mallory proposed moving to Utah so that her children could be closer to their birth father, something they wanted, I relented.

On my last day in the Army, I stood on the roof of the AWG building facing the Installation Flag, listened to a recording of Taps over the loudspeakers, gave one final salute, and loaded up in the moving van. We moved to Utah and bought a house, another Tammyble decision. However, there was one good thing about living in Utah. I spent quality time with my grandfather, the sailor who influenced me to join the military. I would stop at KFC every Sunday, get him his favorite order, and go to his house. He had lived there for almost 60 years, and it was still in pristine condition. I would sit with him for hours. We bonded during those Sunday visits in the same way Wally had shown me. I felt he understood me at a level that few people did.

At this time, I struggled with an identity crisis. After all, I had been in the military for nearly 30 years; I was institutionalized[1]. I had been told what to wear, where to be, what to do, and how to do it. I was part of something greater than myself. On the day I gave that final salute, I had been a soldier who had risen from Private to LTC, had served 27 years, 20 of those in SOF formations, had deployed six times (each one with a Special Operations Unit), had earned several awards and decorations, and wore a triple canopy[2].

I had earned and was given a certain level of respect. The following Monday, I was just "plain old Matt" in cargo shorts and a t-shirt, trying to figure out the civilian world in a town without any military presence or point of reference. I wasn't sure how to navigate this strange new world.

Additionally, I struggled to find a job. I had been diagnosed with suicidal ideation in 2011. It had been a thought in my head for quite some time now. It was fueled by a combination of beliefs that I had spent a considerable amount of time fostering. One was that life was meaningless and lonely. Second was that my struggles were too overwhelming to overcome. And third, was the idea that joy and happiness were out of reach.

For the next eight months, despite living in separate parts of the home, I would be removed from my house over ten times and given a restraining order each time. Not once did I lay a hand on her or her

---

1. The military results in a certain degree of institutionalization.
2. A term used to describe someone with three tabs, a Special Forces Tab, a Ranger tab, and an Airborne tab.

kids—or any other woman for that matter. She was always quick to point out that I was a Green Beret with PTSD, and that was all law enforcement needed to hear to side with her.

The constant fighting with Mallory, the identity crisis of leaving the military, failure to find meaningful and adequate work, constantly being forced from my home without cause, PTSD, and alcohol and narcotic addiction had reached a tipping point. With Veterans Day approaching, I was overwhelmed and full of rage. Naturally, there was a fight between Mallory and me, and I took my gun and drove away. I intended to take my life that weekend. The first night, I drove to a bar called the Notch in Kamas, Utah. It was a small bar with maybe eight tables, a small dance floor, and a stage. There was also an adjacent pool room. I decided to pull in for the night before driving to Moab the next day. I would have a couple of drinks, then pull off on the side of the road somewhere in the Uintah National Forest.

Not long after I started drinking, a reggae band began to set up on the stage. I thought that was peculiar, given the size, location of the bar, and the hard-living town we were in. I fully expected them to be run off by the meager crowd. But not long into their performance, the energetic blonde female singer had the crowd singing along to some Bob Marley covers and some original material.

I was still drowning in my misery, looking for solutions to my God-sized hole, in the proverbial bottom of a glass. Right before their mid-set break, they took a moment to honor veterans of our country and thanked veterans everywhere for their service. Holy shit—I ran out of the bar and sat in my truck, crying like a baby. Maybe I did matter; maybe someone did care. Maybe all those years in Afghanistan and Iraq hadn't gone unnoticed. If a small reggae band from California, putting on a show in a small bar at the edge of a national forest for a crowd of eight, can pause and recognize the contributions of veterans everywhere, maybe there was hope.

It may seem insignificant, but I credit them for saving my life. I returned to the bar, slowed my drinking pace, and listened to their music. I was fixated. Joy seemed to exude from them; they appeared overwhelmingly happy. I was mesmerized. At 2:00 a.m., the bar was closing. The bartender asked me if I was driving. To this day, I am a huge reggae fan.

I don't recall what I told him, but it was enough that he said, "Hey, I also happen to be the bar owner. You're welcome to sleep it off in the parking lot."

I took him up on the offer. I moved my rifle from the back seat to under the seat and curled up on the back seat. It is cold in Utah in November at an elevation of 7,000 feet. But I was drunk and numb enough to sleep through the chill. I woke up to the sun and started driving.

I didn't have a destination; I just drove south. I was looking for someplace warmer. As I drove, I listened to the band's CD I bought the night before I wound up in Moab, Utah. I still had suicide on my mind, but it was fading. I spent the remainder of Veterans Day weekend camping out of my truck and sitting in the desert, looking up at the stars. By January 2018, I would accept the failure of another marriage and move on with my life.

In May, one of my closest friends set me up on a blind date with Sue. Sue had a quirky vibe with serial killer undertones. On our first date, she went off on a tangent about how, during a natural disaster, it would be super easy to kill everyone on her list and dump the body at the epicenter of the disaster. "After all, no one is doing autopsies." Sue lived by the motto of "easy breezy." She became my first medicine. She reminded me what it was like to laugh, something I hadn't done much for over a decade.

Although Sue and I got along very well, I still had the same old problems. On Memorial Day, I told her, "I'm gonna go camping alone for the weekend."

"Okay . . ." she answered with an inquisitive tone.

"Yeah, it's a hard weekend for me, and I just want to be alone." I continued.

"Makes sense." She said supportively.

"I'll probably sit around the fire drinking myself numb." I offered in candid honesty.

"Hmmm." I later learned this was a red flag for her, as it should be. I thought I was showing her I was responsible enough to get away from anyone and take my rage into the mountains.

I thought I had experienced several "rock bottoms." I was about to find out there was a new lower low.

During Father's Day dinner, I had argued with my parents over a silly tree. By August, I decided to try to mend the fence and drive to their home to apologize. However, my ego still needed to be "right" by attempting to make my point again. My stubbornness wouldn't let it go. Before long, the same argument had erupted and escalated. I was directing F-bombs and every other obscenity imaginable at my sweet, religious parents and sister who had come into the room. By then, I noticed the early warning signs of a violent outburst—elevated heart rate, flush, tight fists, rapid breathing. I was in my fight-or-flight, and my default between the two was to fight.

In an attempt to shift to flight, I said, "Fuck off and fuck you all; I am out of here," and started moving toward the door to escape.

Halfway to the door, I heard my father get up from his chair, his feet crossing the hardwood floors, and yell, "Get back here."

Since this event, we have spent time unpacking his words; he wanted an apology to my mother for my language. In less than a split second, my lizard brain went from three options (fight, flight, or freeze) to just one. Freeze had been removed from my choices through 27 years of military service. Flight was just taken away by the "get back here" ultimatum. All I had left was fight.

My father was about 3 feet from me when I balled up my right fist and swung at the wall. It was gonna be my one and only warning shot. My mind was quickly processing my second, third, and fourth moves. Another outcome from years of hand-to-hand combat training. In that moment, I was willing to harm him; there was no question in my mind. My hand connected with the drywall, leaving a hole the size of a cantaloupe.

"Call the cops," Dad said to my sister.

"Go ahead and call them, see if I fucking care." I followed up quickly, standing there with clenched fists, waiting for him to move one step closer.

My mother was crying, more like wailing in a way I'd never heard before. My dad had his eyes locked on me. He wasn't showing any rage or violence; it was not his nature, but he showed determination. Even at 75, he wasn't letting me back down the hall, back into the living room.

After a few silent moments, I said, "I'll wait in the driveway."

I went to my truck and tried to cool off. The cops took 45 minutes to get there. Now I was more pissed at how much time was being wasted, and I just wanted to get on with my day.

The sheriff's deputy arrived, blocked me in the driveway, and approached the truck. "What's going on, sir?" The deputy asked.

I calmly explained my side of the story. There had been a disagreement; I got up to leave, and my dad ordered me back. I punched a wall to keep him from advancing, and he called 911. I explained I'd been sitting here for 45 minutes.

He said he would go into the house and speak with my family, and positioned two other deputies to watch me. He was in there for a long time, and I was growing increasingly impatient by the minute. This made for quite the spectacle in a town of 500 people with no regular law enforcement presence. Cars were now regularly driving past the house and slowing down to gawk. After almost an hour, the deputy returned to my truck.

I rolled down the window, leaned out, and asked, "Can I go now?"

The deputy took a couple more steps toward the truck and placed his hand on the top of his pistol. My training kicked in. I knew this was a sign of the serious nature of what was happening. Until now, I hadn't taken any of their presence seriously. After all, I was used to cops coming to the house, but this was different.

"I need you to step out of the vehicle and place your hands on the hood of the truck."

PART TWO
SHAMANISM

8
GRANDMOTHER

"Ayahuasca loves to take prideful people and rub their nose in it. I mean, it can make you beg for mercy like nothing. You have to really approach it humbly." - Terence McKenna

I had my rights read to me, handcuffed and placed in the back of a sheriff's SUV, and driven 45 miles to the county jail, booked, and locked into my holding cell. I was charged with a class B misdemeanor, "destruction of property during a domestic disturbance."

My cell was small. Concrete walls, steel door, small slot where meals were passed to me, stainless steel sink, and toilet. I felt like a caged animal; I was still in fight-or-flight. I paced three strides, turned, and paced back for hours. I had no information, and there was no urgency on their part. I couldn't even ask anyone questions.

So, there I was, pacing back and forth in an 8'x4' cell for hours. It was what was needed to allow my fight-or-flight response to dissipate. I was angry. I swore never to speak to my family again—and I meant it. I was certain that Sue would break up with me. But two things happened while I was there. One, I realized I was being dishonest. Dishonest with myself mostly, but with everyone else around me, too. Dishonest about the severity of my PTSD, of my rage, of my addictions. Two, I realized I was living out the epitome of crazy—"doing the same things over and over and expecting different results." I needed to get brutally honest and needed new solutions.

When I was finally taken to the deputy in charge for my one phone call, I asked, "How much is bail?"

He casually replied, "$1,000."

"Can I have my wallet, and do you take Visa?" I asked.

"Yes." He responded, stood up, and went into the storage area behind his desk, behind a five-foot-tall counter and glass that extended to the ceiling.

He returned and slid my wallet under the opening. I removed my Visa and paid my bail.

"Slide your wallet back under the glass. I'll return it to your things, and you can have it back when you are released." His tone didn't communicate that he was open to further questions, and I returned to my cell to wait.

Several hours later, my payment was successful, and I was released. I was given several pages of instructions: a no-contact order with my family, when to appear for my preliminary hearing, and an inventory of my personal belongings. Then, I was escorted out of the building. I was in for less than 24 hours.

My first act was arranging an Uber ride to my truck. My second priority was to call Sue and break up with her. I was processing a whole new reality and convinced that no one wanted to be with someone who was charged with domestic violence.

She simply responded, "What's the big deal? Aren't we having fun just hanging out together? It's not like we are going to get married or something."

This would be one of the many "Sue-isms." Simple, down-to-earth, no bullshit, tell it like it is philosophies. We stayed together.

At my hearing, I pled no contest. The prosecutor had offered me a deal of one year's probation and an anger management evaluation, followed by anger management classes, with the promise of expungement at the end of the year.

Shortly after my arrest, Jannice was visiting family in Utah and asked to visit me. I was in that dark place and was seeing how my trauma traced back to my adoption. Later, one of my shamanic mentors would describe this as the "original wound." He taught me that it is always easier to address the original wound than the wounds that grew out of it. For example, my adoption created a fear of abandonment; my fear of abandonment created a need to prove my worth. My efforts to prove my worth led me to the most uber-masculine career imaginable, which led to 40 months of combat, which led to alcoholism, addiction, PTSD, and violence issues. I had this realization several years before I met my mentor, but I knew it was all connected.

I wasn't eager to see my biological mother. I was filled with rage, and my life felt out of control. I warned her that I wasn't in a good place. When we met in 2013, she had promised me truth and transparency. Her work as a psych nurse had informed her of my probable issues long before we ever met. She lived up to her word and offered to listen and be honest. My words were short and to the point at first, quickly transitioning to full-blown yelling filled with vile insults and curse words within minutes. She sat on my couch, afraid to look up, while I unleashed all the blame and frustration of my life for her selfish decision. I was the victim of her selfishness and shortsightedness. Forty-nine years of pent-up aggression finally had an outlet and a target.

My Grandpa, the sailor, had been moved to the VA assisted living center. He was not doing well and was going to pass soon. The absolute worst outcome of my restraining order was having to leave his bedside when my family showed up to visit. Shortly, he passed away, and I would have to get a court order just to attend his funeral. It cast a dark cloud over an already sad event, and I hung out in the back of the chapel, hiding from my aunts, uncles, cousins, and family.

Before I was arrested, I was handed a life sentence. If you Google "How to cure PTSD," this is the exact "cut and paste" from the top search result, and it has been for years: "Post-traumatic stress disorder (PTSD) can be treated with psychotherapy, medications, or a combination of both. A mental health professional can help determine the best treatment plan." In other words, according to conventional wisdom, you're fucked.

But, as you know, I am kind of stubborn and don't like to quit, especially if I don't like the answer. I began to Google, over and over and over again, each time trying new and different combinations of words to find any sliver of a solution.

Alcohol, the self-medication of choice for most of the military, was dangerous. It would work to a degree; I could drink enough Jameson to quiet the demons, and it allowed me to pass out at night without being afraid of sleep. It was especially effective alongside opioids to help me get some sleep. But the consequences were extremely risky. Overdose was a looming danger. Alcoholism was near guaranteed in my case, given my susceptibility to habit and my pre-existing mental health struggles. My liver would not be able to sustain the large amounts of alcohol that my brain insisted would help. Worse yet, alcohol has a way of bringing out violent tendencies in many of us. I did not want to allow alcohol to push me toward hurting those I loved.

I was looking for something that met the following criteria: 1) an actual "cure" for my PTSD, not just treating the symptoms. 2) It wasn't a pharmaceutical. I had taken enough pharmaceuticals, and I didn't want another chemical lobotomy. 3) I didn't need to bring up the past once a week, i.e., talk therapy. Some studies indicate that certain factors, such as trauma re-exposure, excessive rumination, poor therapist fit, and ineffective treatment, can lead to worsening symptoms. I had experienced all of those outcomes.

After my arrest, I began to research everything I could about PTSD. Soon, I began to hear about something called ayahuasca. It was talked about in PTSD groups on social media, and people I knew claimed to know "someone who knew someone who had done it." I started to read and research all that I could about ayahuasca. Ayahuasca is a hallucinogenic tea made from the Banisteriopsis Caapi vine and the Chacruna leaf found in the Amazon basin and administered by shamans, generally in Central and South America.

The retreats involved nightly ceremonies lasting between four and seven days. I learned that the "medicine" causes a person to "purge," usually taking the form of vomiting, sweating, diarrhea, shivering, and crying. I learned that the person taking the ayahuasca would encounter "Grandmother Ayahuasca" and that she would show them what they needed to see, not necessarily what they wanted to see. I learned that, with the guidance of a shaman, a person may be able to cure their PTSD.

I read and watched testimonial after testimonial of happy participants who found relief from their PTSD, including everyone from veterans to sexual assault victims. They all said the same thing: they felt cured. It seemed to be the answer to the question that burned in me. I had to find Grandmother Ayahuasca, and I would spend the next 18 months trying unsuccessfully to find it. I found myself as a fifty-year-old with no drug experience, trying to break into a world that was very foreign to me. With my military bearing, many people in this unfamiliar world assumed I was a narc. Humorously, I assumed every one of them was a narc. I would later learn that there is a belief surrounding ayahuasca that "you don't find her, she finds you."

On the Winter Solstice in 2019, Sue's son, Kyle, passed away. The "sun" in her life had been extinguished; it shattered her world and the world of her entire family. Looking back, I can confidently see why ayahuasca wasn't looking for me yet. We had our hands full for the near future, processing the grief and loss. Had I known then that my path would include ayahuasca, I would have seen that it clearly wasn't the right time. But I didn't know, and I stubbornly kept looking for her. This would become another breadcrumb.

In January 2020, the world plunged into a global pandemic. Despite not finding a suitable option, it was clear that I wouldn't go to South America and face the probability of being stuck there.

In April 2020, I was fired from my first post-military civilian job at the Arbinger Institute. I was never given a clear answer. I believe it was because I refused to apologize to the owner, who criticized my leadership style. It was almost humorous to hear a 30-something millionaire running an inherited business criticize my leadership, shaped and refined in the furnace of fire we call combat. So, no, I wasn't going to apologize to him; he had been in the wrong, and I called him on it.

This is how things operate in the SOF community; rank is more subtle in SOF, and experience and expertise matter more. Here is the ironic part—they teach leadership seminars and sell leadership material, including a course for "how to work with and manage difficult employees." The protocols call for several steps before a firing. Instead of following their material, I was blindsided. They showed their true colors, which would be another breadcrumb and a huge blessing.

The timing wasn't perfect for Sue and me. It was just six months after the death of Kyle. Fortunately, I had been talking to a friend I served with in 3rd GP. They were looking for a guy with my skill set, and I was looking for a change. I accepted the job and moved to North Carolina for the 5th time. We would have to date long-distance.

Ayahuasca was still in the front of my mind, and I kept looking for her. I reached out to a dear friend, Jenny, with whom I had a conversation in 2020 regarding her use of other psychedelics (namely, psilocybin and LSD) to heal from some of her trauma. I wanted to see if she could put me in touch with someone in that community. Jenny spoke very positively to me about her psychedelic use, her trips, and recommendations for ensuring a "good trip," how to integrate, how to set intention, and how psychedelics had specifically helped her heal.

"This sounds amazing," I said to her. "Where and when can we do this?" I threw out dates and locations. I offered to pay her to come and "trip-sit" —stay with me while I took psychedelics and help make sure I didn't have a negative experience. I was considering Denver or another city where psychedelics had been legalized.

She shut me down promptly. "Oh, Matt. When the student is ready, the teacher will appear."

Shocked, I replied, "Huh?"

Jenny laughed, saying, "When you arc rcady, thc medicine will find you."

In the back of my mind, all I could think was, "What the fuck does that mean? I am ready."

I was pissed.

My egotistical, alpha male "solve any problem with force" mentality kicked in.

"I am ready," I protested.

What kind of hippie bullshit is this? Am I not sick enough? Am I not suffering enough, not deserving enough? It felt like an arbitrary barrier between me and the cure I knew I needed. So, I figured I would do it my way. I would fly to Denver and do it myself at the next opportunity.

In early 2021, I read an article in an ex-Mormon Facebook group. The article detailed the author's first ayahuasca journey in Utah. I devoured the article immediately. She described the trauma she carried, her journey with ayahuasca, the path her healing took, and her shaman. Most notable was where she took her journey: Salt Lake City, Utah. I could barely get my mind around those four words. I reasoned that there must have been a mistake; Ayahuasca was only done in South America, right? Ayahuasca is a narcotic, right? How would it have been possible for her to have used the medicine legally in my home state? It read:

In the middle of a late-night winter storm in Salt Lake City, Mother Ayahuasca lay on my mat with me. She showed me that the trauma and anguish my ex and I experienced had squashed my ability to look at him as my first true love. We had damaged each other so severely it prevented me from considering HIS pain in our break-up. She showed me how he felt, left alone and broken-hearted. I wept for him. Mother lifted the shame and grief from my hand and released it, whispering that I had beat myself up enough about the death of my first marriage. I finally felt peace and acceptance . . . Finally, Mother wrapped me in her arms and told me that the hard work done this past year in therapy was paying off in spades—my girls loved me and trusted me again. They came to me for love, comfort, and advice. It made me whole. Mother said that the times together as a revamped Family 2.0 were the golden years we would reminisce about later. Just like that, I watched my mom-guilt/shame/remorse as it stood up and slipped silently out of the room.

This was my first introduction to the underground plant medicine community. Her words spoke to me, and I decided to connect with this woman; I had questions. She was warm and welcoming and answered all of my questions graciously. She informed me that there would potentially be another ayahuasca ceremony in Utah in June or July of 2021. I told her I wanted to attend. At the time, I was working in North Carolina, but I would time my vacation to coincide with this ceremony, as I was planning to visit my family and Sue in Utah anyway.

This ceremony quickly overtook my mind as a top priority. It's difficult to describe the intense pull I felt toward Mother Ayahuasca. This ceremony felt like my only path. I would check in with my contact once a month and ask, "Do they have dates yet?" and her response was always, "No, sorry."

I began feeling frustrated, desperate, and worried that my journey would never happen. I started to accept that, again, ayahuasca would have to wait until I could travel to South America at the end of the seemingly never-ending pandemic. As we came closer to the summer of 2021, Sue and I set aside a couple of weeks at the end of June for a vacation for my birthday. Although the dates were set aside and marked on our calendars, we kept facing obstacles. We could not find a suitable vacation plan. Either the resort we agreed on was booked, the dates didn't align, or the destination required a COVID-19 vaccination that neither of us was willing to get. And so, as the vacation date grew closer, we had free time and no plans. It was then that I heard a voice. It said *Google ayahuasca retreat Utah.* I brushed it off—*No one advertises for a schedule-one narcotic retreat on a public website.* And I heard it again. So, out of pure amusement, I googled it, and to my utter amazement, there was a ceremony from June 24th to June 27th, during the days we had already blocked out for our vacation. I immediately sent an email and found that there were two openings left. I called Sue, elated, and asked how she felt about me using half of our vacation to attend an ayahuasca ceremony.

"If you want to," she replied.

"Are you sure?" I double-checked.

"Of course." She confirmed.

So, I paid my deposit, and in the space of 15 minutes, I was in. I was finally going to partake in a ceremony over my birthday during the Solstice and in the mountains I grew up in (more breadcrumbs). I couldn't have been more excited. Having been raised in the mountains, the stars, constellations, planets, and cosmic events like the Solstice were always important to me. After I booked the ayahuasca retreat, we booked a three-day vacation in Sun Valley, Idaho.

In the month preceding the ayahuasca ceremony, I had three interactions with Beatrice, my shamas, or the female version of a shaman. Until then, I don't think I knew what a shaman was. Our first encounter was in June, while I was on a work trip to Fort Campbell, Kentucky, over Zoom. It did not boost my confidence about the upcoming journey. Beatrice was all business and no-nonsense. She eviscerated me right from the get-go. She dove right in, reading my energy, reading my soul, and reading my tarot cards.

She then asked me, "Why do you want to do ayahuasca?"

I told her that it was mainly because of my combat-related PTSD, my three divorces, my arrest, being fired from my job, the drugs, and the alcohol. I spoke briefly of the fear I felt from knowing the violence I was capable of and my hopes for real transformation, a cure.

She replied by telling me that I needed to prepare to lose everything.

She seemed almost agitated with me during our conversation.

She told me that Grandmother wouldn't tolerate any of my bullshit.

"What bullshit?" I wondered. *I barely even know you.*

She must have sensed my perplexity because she immediately followed with, "I won't tolerate this bullshit narrative, this 'poor me' attitude, blaming everyone else, blaming the system, blaming your PTSD, or making excuses." She looked up from her tarot cards. "It is time to own your mistakes. Own your faults and weaknesses."

If I had imagined a loving and gentle shaman nursing my soul back to health with a sacred medicine from the jungle, that dream was over. This woman was the spiritual version of a drill sergeant and was in complete control. I knew I was in over my head and made a mental note.

She then asked me about my psychiatric diagnoses. I mentioned that all I had been formally diagnosed with was C-PTSD, alcoholism, treatment-resistant depression, and suicidal ideation. She asked about my medications and physical ailments, and I listed off all my medications. She asked if I was on a ketogenic diet. I wondered how she knew.

She said, "Stop keto right now and start eating vegan."

She alternated rapidly between asking me questions, flipping cards, and giving me instructions. She reiterated the importance of going vegan from that moment up until the ceremony. She said it would reset my substance-addicted brain. She taught me that ayahuasca has one of the highest concentrations of dimethyltryptamine (DMT) on Earth. I had no idea what DMT was then, so her comments went over my head. I scratched DMT on my notepad and circled it to remind myself to look it up later.

She also told me, rather nonchalantly, that I would experience death. She had experienced death many times. I didn't know what that meant. Did she mean that I would physically or figuratively die? Would I suffer some other form of death? Would it be temporary? I didn't have time to process this comment or ask questions—I scribbled and continued listening.

She paused momentarily and asked, "What does death mean to you?"

I said, "Death is like turning a page, where one thing ends, and another begins."

If I were really going to die, I didn't feel afraid of it. After all, death overshadowed my entire life, and I had contemplated ending my own life many times.

Beatrice said, "It's my job to take us through death and resurrection. The process results in pure ecstasy. You are correct that death is a transition."

She circled back to DMT and continued, "DMT is a chemical found in the brains of persons who have died. You can produce it yourself."

Beatrice quickly switched tracks back to the subject of death. She kept speaking of "complete death." She told me that there was no wiggle room in the ceremony. If I had reservations or intentions to hold onto part of myself or hold out from her, attending would be a mistake.

What she said next confused me further.

"We will create a relationship with Her." Referring to ayahuasca as "her."

What did that mean? I underlined the word "Her" twice.

Beatrice continued, "You will both love and hate Her simultaneously. She will be able to see your soul. There will be no lying, no hiding, and nowhere to run."

I was feeling a whole host of emotions: fear, anxiety, excitement, and relief. If mysticism had a feeling, this must be it.

Beatrice told me, "There are seeds inside of you trying to bloom."

This is some crazy hippie bullshit, I began to tell myself.

Next, she turned her attention to the "purging" that most experience when interacting with the medicine. She said almost everyone experiences some sort of "purge." This purge can take many forms. It can be vomiting, diarrhea, crying, sweating, or shaking. Vomiting is the most common and could come in varying degrees. There would be less vomiting if we "gave in" to the medicine without resistance.

Then she paused completely, changed her tone, and said, "But you won't experience any of the physical side effects associated with ayahuasca."

"What? Why not?" I interrupted.

She replied matter-of-factly, "You have done ayahuasca many times before, in different lifetimes, and are accustomed to it. It will not affect you as harshly as others."

This is some next-level bullshit, I said to myself and mentally rolled my eyes.

I am sure she sensed it.

We wrapped up this first meeting by discussing additional add-ons and other medicines to try. In addition to ayahuasca and psilocybin, I added kambo, a poison from the tree frog, sananga eye drops, hapé (sacred tobacco), cacao (sacred chocolate), a nude bath facilitated by a priestess, and a San Pedro hike on the morning of the final day. It's always been my nature to "go big or go home."

We finalized other details of the retreat. We would sleep communally and be required to power off phones upon arrival. Very little food would be offered, as we were expected to fast for most of the experience. We would be encouraged to get as much rest as possible and arrive well-rested at the ceremony. Beatrice informed me that I might encounter angels, ancestors, animal spirits, and spirit guides, and that I was to invite them in. She instructed me to write letters to my ancestors and ask them to be present for the ceremony.

Beatrice closed on a somber note. She said that, in addition to those positive entities, I may also encounter demons, insects, aliens, and snakes. She told me that those were my addictions, and I needed to fight them.

Her final statement was, "Matt, you need to know what sovereignty is. You must be prepared to claim your sovereignty when it feels like you are surrounded and there is no hope."

I clicked the "leave meeting" button and closed my laptop. How do you process that volume of information in an hour? I sat in my hotel room, looking over my notes. My attention was pulled repeatedly toward these words: "death," "surrender," "no hiding," "no lying," "Her," "DMT," "resurrection," "ancestors," "demons," and "sovereignty." What had I gotten myself into? It all seemed much more significant than I had anticipated. Ayahuasca was not a simple drink that would cure me as soon as it passed my lips; it was a voyage into the spiritual plane, a meeting with God and with myself, a convergence with ancestors and enemies, a battle with monsters.

In typical Matt fashion, I closed my journal, set down my pen, and said, "Fuck it. It's too late to turn back now."

But I was still left with more questions. I immediately searched for a shaman or mystic person I could consult with. I googled "tarot card readers near me" and zeroed in on one in Nashville. This would be my first foray into tarot or psychic readers. I am still not sure why I chose to do this—probably because she was a full-blooded Cherokee and listed "shaman" on her website. Her name was Lana.

She told me I was an old soul reincarnated many times. I had fought in World War I, where I saved many lives, including women and children in France. She said that she saw a small village and a church. She told me that after my marriage to Sophie, I'd had many romantic relationships but failed to connect with any of them. She said I did some horrible things to these women and had used them sexually. I was searching for authenticity in my romantic relationships.

She spoke about my relationship with Sue. She said that we would eventually marry and have a long, happy relationship and that she knew the distance between Sue and me was frustrating. She said that Sue had a depression in her life that caused her to fluctuate between happiness and deep sadness. The death of a son brought on the depression. She spoke briefly about Kyle (not by name), saying that he'd had some serious troubles in his life and died tragically. She said he is in a much better place now and at peace.

Then she spoke of ladybugs and butterflies surrounding me; she said I have angels and ancestors with me constantly, watching over and protecting me. She said three of my chakras were blocked: the third eye, the voice, and the heart. She said that my heart had been shattered; there had been a woman in my life who upset my universe, attached massive amounts of negativity to me, broke me, and damaged my spirit. She concluded by saying she needed to tell me more, but the messages weren't clear, and she needed to meditate. She asked me to write down my birth date and full name, then said that she would meditate over the weekend and asked me to return that Sunday, free of charge.

I returned on Sunday. I entered her space and took a seat. She said her meditation had been very successful and that she had received the messages she had requested. The mystic continued to tell me about myself. She told me that, normally, I'm very strong and decisive. Once I make up my mind, I'm "all in" and extremely committed, but lately, I have been depressed, confused, and unable to effectively make decisions. I was trying to connect with the universe without success; she was correct.

At the time, I considered myself agnostic and studied Stoic philosophy, searching for some form, order, or meaning in the universe. She told me that before I could seek answers, make decisions, or connect with the universe, I needed to clear my soul of all of the afflictions associated with the negative consequences of toxicity. She said that our bodies are like storage vessels, and I had been storing depression, toxicity, and pain for too long; I needed to clear it before there was room to store positivity, love, health, and the answers to my questions.

She told me that I would live a long, prosperous life. She told me that I carry a deep wisdom gained from my multiple past lives and that people often seek me out for my advice; even strangers will approach me, seeking my wisdom. She said I'm very intuitive and that if I have a gut feeling, I lean on it heavily because I'm attuned to this type of guidance despite not yet having connected with the universe in this lifetime. She told me that I make my best decisions surrounded by water. She repeated that I'm an "old soul" and most comfortable around people older than me; I'm charitable and very generous, and that despite the negativity, pains, trials, and mistakes I've made, the reality deep down is that my soul is good and I'm a good person.

Lana talked about my sleep habits. She said that I have a hard time sleeping. I'm restless and don't get enough sleep. She said that when I dream, I dream in color. My dreams are specific in their meaning and guide me in my life path. She had meditated on my financial inquiry and learned that I would always be okay financially. She encouraged me to go ahead and follow through with my investment. It would be very prosperous. She reiterated that Sue and I would continue to be happy and eventually be married. Lana emphasized that my daughters would connect with the universe and be happy as well.

She was also quick to point out that I still had unresolved damage from when I cheated on Sophie. Lana concluded with something that created more questions than answers. She told me that my true calling was to be a healer and that I have the ability to lead others to happiness.

Until now, everything resonated deeply. This final comment confused me, but I felt better about the upcoming ceremony. The meeting affirmed many feelings I had yet to put a name to and solidified my intention for my ayahuasca experience. However, what did she mean by saying that I was a "healer and that I had ancestors and angels around me?"

It seemed that the deeper I went, the more questions I had. My entire understanding of this world was beginning to unravel. I felt that Lana was on point. I know there is the idea that fortune-tellers or readers speak in such generalities that anything they say could be seen as accurate. But with Lana, I felt she was a gifted tarot reader and had reassured me.

Beatrice reached out to me a second time, asking to schedule another Zoom meeting. I was initially confused since I had been told there would only be one pre-ceremony meeting with the shaman, followed by a group meeting, the ceremony, and then a post-ceremony integration. I decided to follow the mental notes of our first meeting and "be humble, listen, and take notes." I recognized that I was out of my element and made the appointment as she requested.

Beatrice was less harsh in this meeting but still got straight to the point, telling me quickly that I would have an intense experience. She said that only one other person at the ceremony would have an experience as intense as mine. She said I would intuitively recognize what the others were experiencing. Others would have tourist-like experiences, but she and I would undergo some real hard work. Beatrice reiterated that I needed to completely surrender and let go of everything I held onto. She repeated that the real work begins with integration at the end of the ceremony, but the medicine would stay with me and remind me of the lessons it taught.

Beatrice told me that I would get nauseous and physically ill if I chose to ignore the medicine—ignore it long enough, and I would become incapacitated. She instructed me that I would need to follow her every direction and listen closely, and I would find all the answers by turning inward. She taught me that during the ceremony, I would need to ground myself by returning to my mind and checking in with my body. For example, if the medicine showed me a "captivating blue moth playing the flute," I would need to identify my emotion while witnessing it and decide what that vision meant to me.

She cautioned that I needed to be wary of spiritual pride entering my life and influencing my experience. I was confused. What did that have to do with anything? Though I had once been faithful to Mormonism, at this point, I was solidly agnostic. Why would I have any spiritual pride at all? She said that spiritual pride can be an addiction as well.

She asked me, "What is your relationship with money?"

I responded, "It is a dichotomy. I have more than enough for my needs and don't want for much if I live frugally, but I am always in fear of poverty."

She laughed and said, "Well, get your shit together, align with the universe, and you will have great abundance. Every day will be like being on vacation. The angels try their best to guide you, but you must be aligned."

She switched topics and continued, "Boredom is very bad for you, and your partner needs to know that."

I understood the subtle insinuation.

She continued, "The times you get bored are when you engage in self-destructive behavior, like a bird locked in a cage plucking its feathers."

Finally, she said I needed to decide my life's purpose. Once I do that, I must fully engage and pursue that purpose. She told me that I have many wonderful gifts and I'll be able to bless the lives of others with my abundance. With that, she concluded our Zoom call. It was as if Beatrice and Lana had spoken to each other.

I sat in my hotel room and consolidated my notes and preparatory instructions from the two calls with Beatrice and the reading with Lana.

Things I Was Instructed to Avoid:

- Avoid the use of prescription narcotics and/or street drugs
- Avoid the use of antidepressants
- Avoid blaming others
- Avoid smoking any substances
- Avoid alcohol ingestion
- Avoid excessive consumption of media
- Avoid sexual contact and masturbation for seven full days prior to your arrival
- Avoid binging, supporting, or nurturing eating disorders
- Excessive distorted self-talk is destructive to the environment
- Avoid fighting: hitting, violence, drama, arguing, insulting oneself or others
- Avoid all forms of caffeine for three full days before your arrival
- Avoid engaging in self-harm
- Avoid soda
- Avoid pornography

Diet Recommendations:

- Raw, vegan, or fish (no beef, pork, or chicken)
- Forty-eight hours before the Sacred Ceremony, only consume raw fruits, vegetables, rice, and root foods (no fish, beef, chicken, or pork)
- Twenty-four hours before the Sacred Ceremony, only consume raw fruits and vegetables
- Day of Sacred Ceremony, only consume greens and juice

I was left wondering what these instructions had to do with combat, PTSD, and my overall emotional state. It was confusing and felt off-target to me. Every interaction with Beatrice was more bewildering than the last, always shrouded in mystery and mysticism. But I knew that she knew much more than I did about ayahuasca. She seemed to possess a transcendental wisdom and confidence. Still, I wondered: *Is this what all ayahuasca retreats were like? Were other participants being given the same messages? Did the other participants feel as confused as I did?* I was about to find out. The next scheduled call was the group call, and I would see the faces of the other nineteen participants proceeding on this journey with me.

The group call had a much different atmosphere and dynamic. Beatrice led the meeting and was welcoming, excited, and full of positive energy. Clearly, not the stern tone previously directed at me. This final call was to introduce each of us to one another and give us a chance for final questions and answers. I was one of the oldest people on the call and one of six men. I also felt very uncomfortable as everyone else seemed more familiar with the terms and expressions. I would have described this meeting and these fellow journeyers as "new age." For a hard-core conservative, agnostic, retired military male, I was way out of my element and about to find out what an understatement that was.

In the final week leading up to the ceremony, I used my time to prepare my intention statement and answer the questions I had been given.

What did I want from ayahuasca?

What did I hope to accomplish from this ceremony?

I put immense thought into this, writing and rewriting different intention statements. Ultimately, I came up with four things I was hoping to accomplish based on important questions:

- **What does death mean to me?** *Most often, we think of death as the death of the physical body. I believe that there are many types of deaths: physical, spiritual, emotional, and mental. In the context of this ceremony, death is a spiritual* *and emotional transition—the end of one phase and the beginning of another. This death will be my opportunity to leave the self-destructive behaviors behind, to travel to the ends of my soul—the brink—and return better for it.*

- **What does sovereignty mean to me?** *Supreme power and acting with independence. To claim my sovereignty, I have the right, the freedom, and the power to act for myself and to act independently of others. When I confront my demons, I'll claim my sovereignty from them. I'll no longer be ruled by evil, dark, harmful, and self-destructive behaviors.*

- **Intent statement?** *I, Matthew Butler, willingly submit my soul to ceremony and ayahuasca, to my shaman, and to the medicine. I expect to suffer my death—death of mind (poor thoughts), death of soul (poor character), and death of emotion (ego). I will claim my sovereignty over my thoughts, my decisions, my life, my choices, and my past. I will be free from PTSD. I will gain patience, lose anger, learn how to love, and learn how to forgive.* In hindsight, this is an awful intention statement.[1]

- **Invite your ancestors to the ceremony.** I wrote letters to my maternal grandparents, paternal grandparents, and half-brother.

The day of the ceremony had arrived. I followed all of the diet and preparation instructions. I felt mentally ready. The one thing I was missing from the packing list was a bandana. I knew Sue would have one; it was a running joke in our relationship that she always had three of everything. I asked her for a bandana, and she disappeared, returning with one a moment later.

"Here," she said. "This was Kyle's. He would want you to wear it."

"Are you sure?" I asked hesitantly.

---

1. Ideal intention statements are shorter and easily recalled.

"Yes." There was no hesitation in her voice.

I took it and placed it carefully among the rest of my things. Sue was supportive and lighthearted as I prepared to leave.

She said, "Please don't come back too different. Will I have to call you 'Moon Child' or something?"

We laughed as she helped me load the truck. I didn't know what state I would be in after four days and six medicines later, so I asked her to drop me off and pick me up after the ceremony.

We drove ninety minutes from her house to the ceremony site. The mountains would keep us cool in the hot Utah summer. When we arrived, Beatrice came out to greet us. Sue hopped out of the truck to switch to the driver's seat. I grabbed my bags from the back seat, and we kissed. I told her I loved her.

"I love you, too," she replied, and we hugged.

I followed Beatrice, who showed me the cabin I would share with the other participants. I powered off my phone for what I anticipated would be the next ninety-six hours. I was taken upstairs to a room with cots arranged where six of us would share 144 square feet of space. It was cozy, similar to the tight quarters I had experienced in the Army, no big deal. I noticed a bookshelf near my bed and a book about the anatomy of snakes resting on top of the pile. I knew from my research on ayahuasca that "she" was often associated with snakes. Somehow, the book felt like an omen, but I tried not to let the uneasy feeling seep into my subconscious. We had about an hour to get situated, and then we would meet at four o'clock as a group to discuss the ceremony. My stomach felt upset. I couldn't discern whether it was from a lack of food, anxiety, or something else.

Our group was fourteen women and six men. Participants came from Utah, Pennsylvania, Oregon, Washington, Nevada, Illinois, and New Hampshire. It began to rain intermittently as our group gathered on the patio. I noticed a man exit after a conversation with Beatrice, then go inside for a few minutes only to reemerge with all his gear and drive off in his car. He looked angry, and I was happy he left. I didn't need more anger in my life.

The strong hippie vibe was a stark contrast to my small-town upbringing and my military career. Almost everyone dressed in bright, colorful, loose clothing. There were beads, crystals, essential oils, and incense. I began my usual protocol of categorizing and sizing up the attendees. I could tell which participants were experienced, who were new like me, who were confident, and who were insecure. The four co-facilitators, two men and two women, wore yellow t-shirts for easy identification.

One of the young men, Scott, approached me and handed me a black object, two interlocking triangles facing opposite directions.

"Hey, man," he said. "This is a Merkabah. It helps synthesize our energy. I feel like you'll want to hold onto this throughout the ceremony. Please just return it afterward."

I thanked him with uncertainty.

I was out of my element. I was one of the oldest there, the only military person, and it seemed I was the only person in the group who didn't understand crystals, essential oils, oracle decks, and smudge. I decided to be the "gray man." During the sixth week of basic training, my drill sergeant called my name during a mail call. When I came forward to collect my mail, he looked at me and asked, "Who are you?"

"Private Butler, Drill Sergeant," I replied.

He asked, "How come I have never seen you?" I didn't have an answer for him.

I would try blending in, keeping quiet, listening, and learning. The gray man can observe from the shadows and overhear things that someone less invisible would not. It's a very effective tactic when one is in an unfamiliar place or situation and needs to absorb their surroundings as much as possible.

I overheard people talking about the cosmic energy surrounding the event. A younger man named Gary was talking about his use of LSD.

He said, "UFOs had prevented nuclear holocausts on three separate occasions in the past six months."

"Okay, weirdo," I thought and got up and moved to a different bench.

There were many scattered discussions about Pachamama, LSD, ayahuasca, magic mushrooms, and cannabis.[2] Being raised in a staunch Mormon environment and joining the military hadn't given me much of an understanding of drug culture.

Just then, I caught sight of a hummingbird feeder. Sue had six hummingbird feeders in her yard. Hummingbirds are her favorite. Immediately, I felt peace. I reflected on how ayahuasca had found me and that Jenny had been right. I felt more confident. I thought about how our vacation plans had repeatedly fallen through, how I had quit my job and moved home to Utah in those last two months. The summer solstice had transpired a few days before, along with a lunar eclipse. There was a Blood Moon that night, and the rain washed everything clean and cooled the summer heat. It felt significant that my birthday was the following day. It all felt like it had fallen into place, like tumblers all aligning in one grand, synchronized movement.

Beatrice walked us down to the ceremony site and instructed us to

---

2. The word "marijuana" has racist connotations due to its historical use in the U.S. to stigmatize Mexican immigrants and people of color, particularly during the early 20th century. Specifically with a movie called *Refer Madness*. The term "marijuana" was used in fear-based propaganda and the criminalization of the plant. Today, many prefer to use the term "cannabis" to distance from this racist history.

find spots on the ground. I selected a spot far away from the others and found a place under a large tree.

One of the staff told me, "Beatrice needs to be able to see you. She'll be sitting over here." He motioned to an altar nearby. "Try to keep within this circular area."

I looked around and felt like every place was too close to other participants. The only place that offered a little space was directly facing Beatrice, with my back to a cliff. The cliff was at least 120 feet down. The first 30 feet were angled at about 30 degrees, and the last 90 feet were closer to 70 degrees. The military officer in me wanted to see the risk matrix.[3] I contemplated the fact that I was about to take a powerful hallucinogenic in the open woods—at night—seated at the edge of a cliff. Surely, no one thought this through. Was anyone there qualified to anticipate and prevent accidents or trained in first aid? How far was the nearest hospital or even search and rescue? Everyone seemed to have a blind trust, and as the gray man, I decided to keep quiet and followed the shaman's and helpers' lead.

Beatrice made some final remarks. "What you are about to experience will be divine. You are here for the medicine. The medicine has called to you, and you all have answered."

She introduced four of her helpers, Maddie, Mary, Gary, and Scott, and told us that they would take care of any needs we may have, and to seek them out when their help was needed. Beatrice had us return to our cabins to get what we needed, reminding us to grab our mats, blankets, cold and rainy weather gear, water, and chapstick.

---

3. A risk matrix is a tool that NCOs and Officers use to determine the amount of risk involved in a situation, what steps would be taken to mitigate the risk, and to re-evaluate the overall "mitigated risk." High-risk events call for a sign-off from officers in the chain of command to squarely place blame on the planners and leaders, should something horrible happen.

As she advised us to use the restroom and drink some water, she laughed and added, "Puke buckets will be provided."

She straightened, serious again, and said, "The right people are here today. Everyone who is here is here for a reason. Five people have been denied, and one has left because it wasn't the right time. The rest of you are all meant to be here today. It will be cold tonight. Once the ceremony begins, there will be no leaving the circle until the ceremony is over. No exceptions!"

The spiritual drill sergeant had returned. Before I could claim the cliffside spot as mine, Beatrice directed us back to the cabin, reminding us to return no later than five o'clock. We had just over thirty minutes.

I had no wet or cold-weather gear. Utah temperatures had been holding at a hundred degrees, so I hadn't thought to bring any. I went back to the cabin and grabbed two blankets from my cot, my yoga mat, wet wipes, a small bottle of water, and Chapstick. I put on Kyle's red bandana, a black stocking cap, two long-sleeved shirts, and two pairs of socks. I had brought an electrolyte solution, which I passed out to those I shared a room with. I was in a military mindset, doing Pre-Combat Inspections (PCIs). I knew that I could likely be dehydrated and low on electrolytes. I also grabbed the crystals I had been told to use and the Merkabah. I wrote in my journal *See you on the other side* and moved to the deck until more participants were ready to go down together.

I was still playing the role of the gray man and didn't want to be the first to arrive. A small group of us headed down, and I felt uncomfortable again as we came to the clearing. I still couldn't find a spot to sit that felt right. The only place was the spot at the edge of the cliff.

Fuck it, she did say we would all die anyway, I thought to myself. I laid my yoga mat between two tall shrubs, forming a cave-like entrance. Dead in front of me was Beatrice, sitting like a high priestess at an altar where she would perform her ceremonial rituals, and behind me was emptiness. Unbeknownst to me, at the time, there is the notion that every ceremony is a metaphor for life or a particular situation. I was staring straight ahead at her with my back up against a cliff and no way out.

Beatrice gave one last warning. "Medicine is voluntary," she said. "You do not have to do this, and if you would like to leave now, you may."

No one moved. She reminded us that we would stay there until the end, around midnight, and that no one would be allowed to leave. She double-checked that no one had brought a cell phone.

"Remember to work with the medicine, not against it," she admonished.

Next, she lit sage and smudged the circle, each of us.

When she approached me with the smoldering bundle of sage, she whispered, "You're gonna do great."

People were meditating, praying, and practicing breathwork. I looked around at the other participants in their yoga poses and realized I was no longer the gray man; I stood out.

Beatrice scanned the group as if looking to the spirit to direct her to the person who should go first. She called for Emma to come up and take the medicine. I watched Emma rise from my right side and approach the altar, have a few words with the shaman, and drink the medicine before returning to her place. Beatrice scanned the group again in a clockwise motion.

When I felt that her eyes were about to make contact with mine, she skipped over me and scanned the rest of the group before looking back at me and calling, "Matt!"

I approached the altar and kneeled.

She asked, "Do you have your intention?"

I nodded and took a moment to repeat it in my mind. I stumbled. There was a part I couldn't fully remember. I paused, thought for a moment, and recalled my intention fully. I repeated it in my mind and then looked up at the Beatrice. "Take your time," she said gently.

After a moment, she stretched out her hand with the cup of ayahuasca. As I reached for it, she withdrew it, causing me to look her in the eye, and said, "There's no turning back now."

She pressed the cup of medicine into my hands. I took it with both hands and looked down into the rich, dark brown liquid. I knew that for the next six hours, there would be nothing I could do to lessen the effects of the medicine. What I didn't understand then was that she had meant that my entire idea of reality would soon be changed permanently, that my old life was about to die, and that there would be no going back to the mindset I had before that moment.

I drank the medicine, about fifty milliliters in total, in one gulp. It was a nasty, bitter, vinegary drink, gritty and pungent. There was nothing pleasant about the taste. Even as I write this, the memory of the taste returns, causing my tongue to swell and my throat to tighten. I returned to my seat and sat cross-legged, watching the others as they approached Beatrice's altar one by one. I estimated that close to thirty minutes had passed from the start. Soon, some of the participants began to moan and cry. The medicine was taking effect with the others, but I still felt nothing. Not long after that, others began to vomit uncontrollably.

I wasn't nauseous or queasy. All I felt was anger begin to manifest in my gut, and I thought it was all bullshit. I wanted my money back. Another failed attempt to get out of the PTSD prison that has held me captive. In a typical Matt moment, I decided to challenge the medicine, the same way I challenged God with an axe during a lightning storm.

I spoke to her in my mind, saying, "Come on, is this all you've got? What are you waiting for? I dare you. I fucking dare you to take me to my death. I'm waiting!"

Still nothing. I tried a new approach. I decided to try a softer approach.

I mimicked what I saw the others doing, sitting yoga style, and again I spoke, "Mother Aya, please come to me, please help me. Mother Aya, please help me cure my PTSD; please bless me with your presence."

Still nothing.

At approximately the one-hour mark, Scott approached me and informed me that the shaman wanted to give me a second dose. I readily agreed and returned to the altar. I kneeled for a second time and was given another fifty milliliters. I drank it down, reacquainting myself with the bitter, burned taste of the tea and the grittiness that lingered on my tongue.

This better work, I thought to myself as I returned to my spot.

A few minutes later, I began to feel a warmth in my forehead, and a euphoric feeling crept up from the small of my back. The warmth and euphoria began to envelop me, unlike any feeling I had ever experienced. I was losing my vision to a dark, black void. I began to be surrounded by darkness, a thick, tar-like darkness that encircled me until everything was completely black. Out of fear, I immediately began to feel that I wanted to lie down to avoid trying to walk or move, knowing that I was dangerously close to a sheer cliff. I pulled the blankets over my head and curled up into the fetal position.

I was completely aware of my surroundings. Then, suddenly, out of nowhere, I was met by a face I recognized. It was Kyle.

I let out a loud yell. "Why?" I asked him, intent on learning why he was there.

He greeted me with a smile and said, "I volunteered to guide you through the veil."

He said that now that his mission was over, he would be able to visit his mother more often.

Soon, colors filled my mind. Some I had seen before, but others felt new and unfamiliar. They were layered with geometric shapes of every kind, appearing and disappearing. As I lay there, basking in the colors and enjoying the euphoria, I was mesmerized by these colors and shapes. The colors took on a pearlescent or metallic sheen, with colors shifting as if they were illuminated from within. The shapes were intricate, repetitive geometric shapes, fractals, and mandalas. What I witnessed began to ripple or wave, as if reality was breathing in sync with my breath and heartbeat. The edges began to blur and seem to melt into one another, giving a more fluid and interconnected feel. I was having what felt like the best dream of my life, and I still didn't have the urge to vomit or any nausea.

The shapes continued to float inward and outward; some grew larger and then reversed direction, shrank back into the darkness, and were replaced by others. They seemed to have more than three dimensions as they floated in and out of the center of my mind's eye. I enjoyed the light show. If that was what this was, it was okay with me. I was happy to enjoy it for the next six hours, and I settled in, but the colors soon faded after what I perceived was about fifteen minutes. I sat up and took the deepest breath of my life, what I later called "the breath of life," as though I was a newborn baby drawing my first gasp of air.

This left me in what I now call the "in-between place." I was disoriented and confused. Part of my mind felt like it was still in that other dimension full of indescribable colors and geometry, and part of my mind recognized familiar faces and the landscape. I was having trouble breathing and reminded myself to breathe manually, as the shaman had taught me—deep breaths in through the nose and out through the mouth. I felt suddenly defensive and confused. My hands curled into fists as if to fight anyone who would approach me.

After five or six deep breaths, I knew where I was, and I could remember why I was there. I was left feeling euphoric once more, surrounded by warmth and love. My entire body was buzzing, and I could feel the activity in every cell. I removed the blankets, and the first thing I saw was hummingbirds. There were dozens of them flying overhead and skimming the ground. I had never seen so many hummingbirds in my life. It was another sign that I was safe, and this was an absolutely sacred event. I noticed that my feet were burning up. It wasn't new—my feet are always hot, something I attribute to a lifetime spent in combat boots. The heat kept growing, and when I couldn't take it anymore, I kicked off my shoes and socks, planted my bare feet on the ground, and sat there looking at the colors. Nature was beaming with the most intense colors I'd ever seen. It was as though everything was in ultra-high definition and gleaming. Just then, I noticed Beatrice sitting directly in front of me.

"How are you doing?" she asked.

All I could do was grunt. Most of my words were single syllables through gritted teeth. My manner of communication consisted of deep-throated noises, like some prehistoric language. My mouth formed a frown, my head began to lean to the right, and my chin lowered into my chest so that when I looked at Beatrice, I was looking at her from beneath my brow with my head cocked to the right side.

"Tell me what's going on," she said with a kind smile.

"Don't know." I began to make a fist with my right hand.

"What is that?" she asked, gesturing to my fist.

"This is me," I grunted.

"Tell me more." She pried.

"This is me," I repeated, pointing to my right hand with my left index finger. "This is a weapon." I made the trigger finger motion.

"This is for fighting." I circled my finger around my clenched fist.

"This is destruction." I shook my fist.

"This is my tool," I said through gritted teeth.

"This is death," I said with resolution.

"This is everything." This time with deep emphasis.

We sat silently for a moment as she waited for me to push out my words.

"They were wrong," I said.

"Who was wrong?" she whispered very gently.

"Everyone. Everyone was wrong." I answered.

"What were they wrong about? Please tell me," she said in an inviting and curious tone.

"My birth mother, who left me. The kids who teased me. The town that treated me as an outcast. Everyone was wrong; everyone who ever fucking doubted me was wrong."

"How were they wrong?" she pressed.

"See this fist?" I grunted.

"This fist is me. I won. I always win. I never lose. I am a goddamn Green Beret. I was the best of the best. I was deployed for over forty months. They should never have bet against me. Everyone bet against me, and I won." My tone increased as I pounded on my chest with my right hand over my heart.

I began yelling and pounding, "I fucking won! I fucking won!"

Over and over, I yelled, beating my chest as hard as I could.

Beatrice calmed me. "It's okay, it's okay, it's okay. Did they hurt you?"

"Yes," I said, beginning to cry.

She gestured to my tightly held fist. "Do you know how that makes me feel?"

"No," I said in this guttural tone, almost not recognizing my voice.

"It makes me feel sick," she said. "I feel ill when I see that, and I see that it's making you sick too."

I tilted my head and looked at her with inquisitiveness.

"I'll never give it up," I said, gritting my teeth and glaring intensely into her eyes.

I clutched my right fist with my left hand and pulled it close to my chest.

Scott had been nearby, watching over the interaction between Beatrice and me. She had asked him, among others, to keep an eye on me. She had told her staff that my experience would be the most intense of the group. Scott walked over and handed me the puke bucket that had been resting behind me.

"Hey, brother," he said, "here's your puke bucket in case you need it."

I was already angry, and the interruption antagonized my rage. Without breaking eye contact with Beatrice, I snatched the bucket from his hands and threw it over the cliff behind me.

"I don't need no fucking bucket," I said.

"Well, that makes Mother Earth sad," Beatrice said softly.

I hung my head. "I know. I'm sorry. I'll get it tomorrow, I promise." I whispered.

The shaman placed her left hand on my fist and lowered it to my side. She smiled the biggest, warmest smile. Her pupils were shining like diamonds. Usually brown, they had turned white and blue and were gently rotating like the geometry I had seen when I crossed through the veil with Kyle. Her shape began to shift. I saw Beatrice transform into a panda, a wolf, and a horse, to name a few. My mind was stuck as I tried to process what I saw: this beautiful shaman with diamond eyes shapeshifting from one body to another.

"Concentrate, Matt," she said.

Then she leaned close, maybe six inches from my face, and whispered, "Matt. The universe brought you here. This was all orchestrated for you. Matt, today, you are being called to be a shaman. You have a mission to go forth and rescue your tribe. Matt, do you understand what I'm saying to you?"

I grunted in a tone that suggested that I did not.

"Look around," she said.

For the first time since my journey had begun, I looked around the space at the others. Several people were watching me calmly and smiling at me. I looked back at Beatrice and grunted again, raising my eyebrows as if inviting more information.

"Matt, you are a shaman," she said again.

"Matt, this was all orchestrated by the universe. You were brought here for a reason."

I wasn't processing it. I heard the words and vaguely understood what she was saying, but why? What was going on? I couldn't make sense of it.

Beatrice reached out and touched my right fist again. "Do you remember this?"

"Yes. I still don't want to let go of it. It's everything to me. It is me. I'll never let go of it," I said.

"You don't have to," she said gently.

I looked at her quizzically. She reached out and grabbed my left hand. She carried a certain air of authority and confidence; she had no fear of me.

She continued. "Matt, now, we add the healer."

She brought both of my hands together. "We will honor and respect the ego for its amazing job protecting and keeping you safe. Now, we will center and balance the ego by bringing the healer out."

She gently shook my right fist up and down and said, "Matt, this person can reach people that I can't. You speak their language; they are your tribe. You must go to them. This," she said, glancing at my right fist, "is your credentials. You mustn't lose your credentials. It's okay. Being the warrior is okay; you don't have to get rid of that persona. We're just going to balance it with the healer." She pressed my hands together. "Do you understand?"

"I think so," I replied.

"Okay. We'll work on it some other time."

Just then, one of the other participants began moaning loudly, catching Beatrice's attention.

She turned back to me briefly and said, "Matt, I need to check on my other guests. Are you going to be okay?"

I told her I would, and she left.

I sat in silence, trying to make sense of everything. I brought myself back to my intent. Did I learn how to love? Did I gain patience? Did I cure my PTSD? Did I learn how to forgive? It didn't feel like it. I was overwhelmed with the "plot twist."

I wanted to move, so I asked Scott if I could sit under the tree shading the altar. I made my way to the tree and noticed a stump where a tree had been recently cut. I saw the sadness in that tree. The droplets of sap oozing out of the wood were tears she had cried from being cut down. In that instant, I loved that tree as if she were my child. I kneeled in front of her, placed my hands gently upon the injured surface of the stump, and cried for her for several minutes.

Gary, the young man I'd overheard earlier talking about aliens preventing nuclear war, came to check on me. I had initially dismissed him as a nutjob, but now I literally recognized him. It was the same way you'd recognize a childhood friend after decades; they look different but somehow the same, and you intuitively know them as your friend.

I looked up at him and said, "I know you. Your name is Loki."

He grinned hugely and said, "I know you, too."

"We sailed together. You were my spiritual advisor. We were Vikings. I sailed the oceans, and you sailed the skies." I said, hardly believing the words coming from my own mouth.

He chuckled and nodded. I wasn't sure if he was earnestly agreeing with me or just humoring me and making a mental note to avoid me. Gary helped me up, and I moved to the tree shading the altar.

As I sat down, I connected with the earth. I could smell and taste the rich, dark soil that had been composting and housing life for centuries. Intuitively, I knew that I needed to ground myself in the earth at that moment. I sat cross-legged under that tree and covered myself from the waist down in the soil. It felt so good and restored my energy. I enjoyed the view. Directly across from me was the spot I had chosen, the cliff's edge, and an expansive open landscape stretching for miles as the sun began to set. I was in a good place. I felt at peace. I felt connected to the earth. I felt healthy and alive.

I am not sure how long I sat there before I saw the words "GO LEARN" flash boldly in black and white in my mind.

"What?" I asked.

"GO LEARN," it flashed again.

I squinted my eyes as I tried to understand.

"GO LEARN!" flashed for a third time.

I looked around and saw Beatrice still kneeling beside Thomas. I stood and walked over, trying not to disturb their interaction. As I approached Beatrice, her eyes met mine.

I leaned forward and mouthed, "I was told to come learn."

She motioned for me to kneel beside her, squatting with my weight over my feet. I said nothing, just listened, and watched as she reassured Thomas, who was curled tightly in the fetal position on the dirt. She was questioning him gently, similar to how she had questioned me. I watched as she wisely guided him, as she did me. She gave no answers and only encouraged us to talk through what we were experiencing. I intuitively understood that she possessed the experience of many lifetimes and wisdom beyond conventional learning.

She continued to gently guide him through death. As she had said, we would all experience death during this ceremony, and it appeared as though Thomas was, in fact, dying. Beatrice lightly rubbed his right shoulder and arm and spoke soothing words. After a while, she and I made eye contact and connected telepathically. We were able to communicate with each other with a single glance.

I told her with my eyes, "I want to try."

She motioned again with her hand, suggesting that I take over, and stood, allowing me to move closer to Thomas without a word spoken between us.

"Hi, Thomas. It's Matt," I began quietly.

"Hi, Matt." His voice was barely a whisper.

"How you doing, buddy?" I asked.

"Not good, man. I think I'm going to die." He continued.

"I'm freezing," he said fearfully.

I made a note that he was covered in two blankets, and I reached down to feel for his pulse. As I touched his wrist, I saw "it." "It" was the source of his trauma.

I whispered, "Is that you, Thomas?"

"Yes," he replied.

I was in his journey somehow.

"How old are you, four or five?" I asked

"Yes," he said.

"Is that when it happened?" I inquired. "It" was what he had suffered and something I didn't want to speak out loud.

"Yes," he said softly.

"Go to him, Thomas, go to the four-year-old version of yourself and hug him, tell him that it will be okay. Tell him that you will protect him. Tell him that you will never leave him." I suggested.

We sat there in silence until the shivering ceased. After a moment, I got up, walked back to the tree, sat under it, and pondered what just happened for the remainder of the evening. What was going on? Seriously, what the fuck is happening? Was I just imagining things? Was I hallucinating? Is this what drugs do to your mind? Was I losing my mind?

As the night wore on, Beatrice began to release everyone back to the cabins. The participants began to leave, one by one, or in small groups. Some of the participants were too weak or unsteady to walk on their own. I joined the co-facilitators in assisting the other participants to their feet. Scott and I walked beside those struggling, one of us under each arm, and escorted them up the hill.

At one point, I looked at Scott and said, "I see a conquistador. You were a conquistador in a previous lifetime."

"That's interesting, 'cause I have a Guatemalan bloodline and likely conquistador ancestors." He replied.

I wasn't sure what was happening to me; why was I seeing these things? Even more strangely, why did I feel bold enough to share them? And how is it possible that they were accurate? It was as if I was seeing people for who they truly were—not their physical form, but the most dominant form their soul had taken across every lifetime.

We repeated the trek between the ceremony site and the cabins until everyone was safely up the hill. Beatrice had stayed behind, collecting her things from the altar. I approached to check in with her one last time.

She said, "Thanks for helping, Jeff."

"Who's Jeff?" I asked.

"Oh, you know what I mean," Beatrice laughed. "You know what the problem is? You just aren't a 'Matt' to me. You need a new name."

She turned to leave and asked, "You coming?"

"No," I said. "I want to sit here a little longer if you'll let me."

"Yes, but don't stay too long," she warned.

I returned, sat underneath the large tree at the altar, and covered myself once again with the dirt, taking hefty scoops in my right hand and letting the particles sift through my fingers and fall into the palm of my left hand. I looked out again over the landscape, taking in the valley expanding in front of me. I felt euphoric, tired, energetic, and overwhelmed simultaneously. Most importantly, I felt a deep, indescribable love for everything. I felt love for my fellow man, my shaman, the medicine, the journey, Earth, and myself. Just then, I had a thought.

I had asked Sue to marry me for years, but she always declined. I knew she was still mourning Kyle and needed time. I realized then that I knew what it meant to love completely, purely, and deeply. I hadn't asked her to marry me out of love but out of my fear of abandonment. This was, perhaps, what had led to so many of my troubled relationships in the past. For the first time in my life, I saw *my* self-worth. I sat enraptured, reflecting on how I was a soul with infinite worth and perfect as is.

There is only one of me. No one could ever be me better than I could be myself. In this moment of self-love, I realized that even if things between Sue and me didn't work out, I would be okay, with or without her, or any woman, for that matter. I didn't need a woman to complete me. But I had never loved Sue more than I did right then.

I thought of Beatrice telling me I needed a new name. What is my new name? I thought. Do all shamans need a new name? I stared at the moon, which looked so close that I could reach out and touch it. It was known as a Strawberry Moon as it appeared full for about three days, from Wednesday morning to Saturday morning. I thought of Sue. She'd already given me my new name: "Moon Child."

"I am Moon Child," I said aloud, and my body shivered with chills.

I walked barefoot back up the hill to the cabin. Everyone else was asleep when I entered the shared room, but I lay awake the entire night. The effects of the medicine had worn off. However, my mind was still buzzing with activity, sorting out what had transpired over the past several hours, coming to terms with my newfound identity as a healer. I pulled out my journal and made a list of almost thirty people I owed an apology to. This was the first step in making reparations with the other people in my life. Many thoughts passed through my head, some lingering, some visiting temporarily before disappearing. I was given many visions. One of the visions I witnessed involved me standing on a mountain surrounded by juniper trees. I was wielding a large wooden staff, reminiscent of Gandalf's, with a twisted mass of roots at the top.

As the night wore on, my ego returned. The prevailing thought I had was that I was fucking crazy. It was my ego's job to protect me, and thus, it began convincing me that "I was no shaman." I felt embarrassed and reminded myself of the gray man promise. Everyone in the group had seen me, heard me, or even been carried to the top of a hill by me. Now, I am under the delusion that I am a shaman named Moon Child. Seriously, *what the fuck?* I thought. When I noticed that the sun was beginning to rise, any hope for sleep was gone. I recommitted to my original plan to keep my mouth shut and stay in the shadows. This wasn't about me; I needed to stay in my lane for the second ceremony.

Day Two, June 25th, My Birthday

I lay pensively on my cot, listening to the sounds of sleep and deep night creeping into the early morning; I hadn't slept at all. As the sun rose, I decided to get out of bed and stood staring out the window. I was surprised to see a flock of turkeys strutting on the opposite side of the glass. They stared back, unafraid, meandering from the cabin and into the woods. Watching them, I felt lucky to see and connect again with Mother Nature.

I slowly got dressed while the others slept. I had some time before everyone else would be awake and begin the day's activities, starting with a group prayer at eight o'clock. I remembered the puke bucket I had carelessly tossed the day before and decided to use this time to fetch it.

I walked alone back to the ceremony site, still barefoot. When I reached the cliff, I put on my shoes to protect my feet from the jagged rocks and began to lower myself backward over the cliff's edge, clutching to trees and shrubs so I wouldn't tumble to the bottom. I spotted the bucket farther down and carefully worked my way toward it. When I was about five feet away, I maneuvered my body ninety degrees to the right, reaching for it. As I was about to grab the bucket, my right foot dislodged a grapefruit-sized rock, which rolled directly into the bucket. The bucket turned over, picked up momentum, and went straight over the lip of the cliff. It was at the bottom of the cliff in a matter of seconds.

"Fuck!" I muttered.

I grasped onto large rocks and trees to pull myself back up the cliffside. I had promised to retrieve the bucket, reminding myself of my commitment. I stood and surveyed the terrain, planning my route to the bottom of the ravine. I would have to backtrack in the opposite direction, back up the hill to where the ravine began, and follow it down until it met the bottom of the cliff.

As I hiked, I was thinking about the significance of the bucket. I was being taught a lesson. Throwing the bucket was a metaphor for all the times in my life I had lost my temper. Sure, I felt powerful in those moments of anger, but ultimately, I created pain, trouble, and additional work for myself. Thirty minutes later, the bucket was in sight once more. It was lying on a pile of dead tree limbs that had gathered at the bottom of the cliff through snow and rain.

As I picked up the bucket, I saw the most amazing turkey tail feather lying next to a large, dead tree limb perched atop the pile. It was exactly like the one I had seen in my vision, down to every knot and gnarled branch. I picked them up and laughed to myself. My ego mind was battling this new reality.

Okay, universe, I get it, I thought. It wasn't a dream. It wasn't just a hallucination. It was real. It was all real, and I'll stop denying it or trying to talk myself out of it now. I collected my staff from the branch pile and used it to help me back up the hill. But I later learned that the battle between my rational ego and the universe would rage on for a long time.

Day Two: You are reborn, Moon Child. I have no words. Really, it seems like what I was given, what I experienced, could never be lost. Whether I write it down or not. It is so clearly imprinted in my mind that I will never forget it. But it will be revealed to me in gems that I will use when I need them. I need to honor and respect these gifts. I also need to focus on humility. She isn't going to bless me—or, more importantly, my mission—if I am arrogant and full of pride.

Journal entry

After morning prayers, I replayed the events in my head. I fluctuated between trying to accept that I might actually be a shaman and my inner desire to be a "quiet professional." When it was time for my cacao bath, I was told to change my clothes and wait on a bench near the front of the cabin. I sat on the bench, watching the hummingbirds flutter, waiting for my turn for the palm reading. All of this was still so new to me. In a matter of months, I'd had a palm reading, four tarot card readings, and was diving headfirst into plant medicines. My head was struggling to reconcile it. I'd only wanted some relief from the PTSD, and I was beginning to feel swallowed up.

I debated whether this was normal for an ayahuasca retreat. As I debated, a woman named Maureen walked past me; spiritually, I recognized her. She was Celtic royalty, a queen. I watched her for a few minutes out of the corner of my eye and waited for an opportunity to speak with her.

I approached and asked, "May I speak to you?"

She said, "Yes."

I sat down with her on the deck.

"I'm not sure how to say this, but I see you as a Celtic queen." I started

She looked at me with dismissal.

"Ah . . . okay. That's interesting," she said. "Well, thank you very much."

I wanted to dive into a hole. I thought the plan was to keep your mouth shut.

I was saved by the palm reader inviting me to begin. She gestured to a reclining chair and asked me to get comfortable.

She started with my left hand:

Emotional in nature, line of the heart

Marriage on the horizon

Intellectual strength

Fate line: stable

Suited for a life of physical labor

Continuing to my right hand, she recorded:

Affectionate

Highly intuitive

Many past lives

Mystic's cross

Many spirit guides

I wrestled to make sense of what was beginning to feel "right" to me and my ego, convincing me that I was crazy.

That night would be much like the first night, where the same instructions were repeated. Once we were all gathered, Beatrice followed the same pattern as the previous night. She prayed, and then she smudged herself, the ceremony area, each one of the participants, and the medicine. She smoked sacred tobacco. Next, Beatrice poured the ayahuasca and called the participants forward.

When I was called to the altar, I kept my original intention. I repeated it in my mind, head bowed, and reached for the cup full of sacred medicine. I again downed the bitter liquid in a single gulp and returned to my spot. I had less anxiety now that I knew what to expect.

Mother Ayahuasca didn't waste any time that night. I felt the familiar, warm euphoria building deep in my lower back and working its way up my torso and over my shoulders. I greeted her like an old friend, but held anxiety about what I might see in my subconscious. The feeling grew more intense as I felt Her course through my body and into my brain. Part of me expected to see Kyle again, as I had thought he would always be my spirit guide, but he never showed himself. Instead, I would meet others who had much more to teach me.

The geometry and colors returned as my mind witnessed the most amazing display. The shapes burst and shrank, rotated and inflated. The supernatural colors were more than could be imagined. Then, suddenly, the blackness returned.

"What do you want me to see?" I asked her, squinting my closed eyes.

Finally, a red, serpentine line, very small at first, began to emerge from the blackness. I leaned forward as if to zoom in on the line.

What is that? A snake? I wondered.

I kept leaning forward, training my eyes on the shape until it came into focus. It was a line of humans, all men. The line stretched so far into the horizon of my mind's eye that I could only see the end nearest to me.

"Who were they?" I didn't recognize anyone.

I brushed the line with my speed to the last person in line; perhaps I would recognize him. I swiped faster and faster, moving the end of the line closer and closer. The line ended with a sudden halt. There stood a giant of a man. He was easily 6'6" and approaching 300 pounds. He had long, gray hair and an impressive, braided beard. He carried a battle-scarred ax and wore animal skins.

"Who are you?" I asked.

"The first," he said.

I knew immediately that he was an ancestor. I instantly processed that the serpentine line represented generational trauma that had been passed through every ancestor, starting with this man and culminating in me. I would later learn his name is Je.[4]

"Ah. It was you!" I said, laying the blame at his feet. I formed a fist and began beating brutally against my chest, just as I had the night before.

I yelled, "It started with you—it ends with me!" in a rhythmic cadence.

I was unaware of the discomfort I was causing the others in the circle. As I yelled, he lunged forward intimidatingly as if to scare or silence me.

"No," I said. "I'm not afraid of you."

He stepped back with an expression of surprise, unaccustomed to having someone stand their ground against his advances.

This ends now! I shouted in my head.

---

4. Pronounced "Jay"

His shoulders slumped, and he lowered his raised ax. He turned to walk away from me.

"Where are you going?" I yelled. "Get back here! I'm not going anywhere."

He continued to walk away.

"I am not leaving until we resolve this," I continued.

His head hung low as he stopped in his tracks, still facing away from me.

"I am not going anywhere," I said again, softer but firm.

Slowly, he turned and faced me. In a split second, I saw his soul. The massive, imposing warrior figure was now more than those parts. I saw frailty, vulnerability, and humanity. I saw a father who loved his children and had tried his best to raise them. I saw a husband who loved his wife despite his mistakes. I saw a man dedicated to protecting his village. I saw a man haunted by his demons, a tired man who carried a burden for his lifetime and beyond. I saw a man who had done his best but suffered from his own insecurities, weaknesses, and shortcomings.

I saw myself. I reached out my arms and wrapped them around him, hugging him gently. In that instant, the thin serpentine line of ancestral warriors raced back to the beginning. At one end was Je, and now suddenly, at the other end, there a face that I recognized; it was Grandpa. The man who influenced me to join the military, who I idealized, who never spoke about the war. I saw his soul. I saw his frailty, vulnerability, and humanity. Another man who, despite trying his best to be a father and husband, had his own weaknesses and shortcomings, his own PTSD, which he had passed along to my mother, who in turn had passed it along to me. I cried and hugged him and held him.

After a few short moments, I drew the "breath of life" and was again caught in the "in-between place." After some deep, cleansing breaths and centering myself, I came back to reality and recognized where I was once more in the circle. Seeing the others watching me, I realized I had failed again at being the gray man.

Day Three

Day three was the psilocybin ceremony. Before then, people still filtered through the kambo, cacao, and palm reading stations. During these downtimes, I still saw the past lives, spirit animals, and spirit guides of each guest. Simply put, my third eye was working overtime, and I couldn't shut it off. I randomly approached others and shared what I saw in the most respectful manner I could. Nearly everyone was receptive and affirming of what I saw. Maureen was the only one who was very enthusiastic. Midday, she bumped into me.

"I am so glad I found you." She said enthusiastically.

"Oh, really?" I said, shocked.

"Yeah. I just got my palm read." She informed me.

"Oh, cool," I said.

"Can I show you what I was shown?" She asked.

We moved over to a table and sat down.

She held her left hand out and said, "Do you see this shape on my middle finger?" She pointed.

"Yep," I confirmed.

She swapped hands, now pointing at her right hand. "And do you see this identical one on this middle finger?" She continued.

"Yep," I said, following her demonstration.

"Do you know what that means in palm reading?" she asked.

"No clue," I said flatly.

"It is called a 'Rajalu,' and it means that I have a past life as royalty." She said with a grin.

"Woah," I said as chills ran up my spine.

"She said that less than 1% of people have two of them." She added.

I was speechless.

"I just wanted to thank you for being courageous enough to tell me; they both really confirm the validity of each other separately. Thank you." She said, hugging me.

Tonight's ceremony was going to be a much less informal event. Beatrice gathered us together and asked us to sit on the benches that formed a circle. As we waited, there were discussions about who had and had not tried psilocybin, what to expect, and an eager energy. Some complained about the taste; others asked how they would be served or what strain we would be given.

I sat there in amazement, shocked by the amount of experience some of the group had, or even the questions that the other newbies like me had. I knew nothing about magic mushrooms. The mood felt closer to kids about to be let out for recess. By now, I had seen almost everyone's past life or spirit animal. Beatrice instructed us, saying that she would come around and give us our medicine and that we were not to wander too far but to enjoy nature, dance, or meditate.

Soon, she approached everyone individually and reached into a cloth bag with a woven pattern. Holding our hands, she would reach into her bag, pull dried-up mushrooms out, and place them into our hands. She approached me, pulled out two small, dried mushrooms, and placed them into my hand. Then she looked at me, paused momentarily, made a contemplative expression, and reached into her bag and pulled out another. She closed my hand and said to enjoy myself. I chewed on the dried, earthy, somewhat pungent mushroom, unaffected by the taste, and waited.

I wandered around the area, waiting for something to happen; very little did. I did feel very connected to Mother Earth. I eventually found an empty hammock and lay down, staring up at the clouds. I saw soft pastel colors in the sky and heard Kyle's voice.

"Now, do you understand?" he said.

"I think so," I replied.

"This was my escape." He continued.

"I get it," I said with sincerity.

"I only used the medicines and went into nature." He concluded.

I nodded with tears rolling down my cheeks.

After they began to wear off, I wandered over to the deck and watched a small group of fellow guests dancing in a circle. They were moving their bodies slowly and smoothly. There was one woman who was the only one I hadn't been able to see spiritually.

Watching her dance, I heard, "She is the divine feminine."

"Ahh," I said.

When the dance ended, I asked gently if I could share what I saw. As I did, she began to panic; I recognized a level of fear that I had seen many times before in some of the most horrific circumstances. She was dropping into a full-blown panic attack. Other guests rushed to us, and others searched for Beatrice.

Soon, Beatrice joined us, asking, "What happened?"

I explained that I asked to speak to her and that she started having a panic attack. Beatrice put her arms around her and took her inside. Soon, I was sitting all alone and feeling incredibly insecure. My ego shifted my internal dialogue from *you are crazy* to *you are a monster.*

I had no counterargument. I began walking down the road. My anger and rage were surfacing. I thought ayahuasca was going to change me; I thought she was going to take away my demons. I thought I was going to be filled with love. If my very presence has the power to trigger someone into a panic attack, surely this was bullshit. I walked for at least an hour before returning. I found myself sitting at the altar, under the tree. Soon, Beatrice found me.

"What are you doing?" She asked.

"Waiting for you," I replied.

"Oh yeah?" she asked inquisitively. "What for?"

"Well, if she was that deeply triggered by my very presence, then surely this didn't work, and I want the demons out of me," I said flatly.

"That's not how it works, Matt." She said.

"I want more ayahuasca," I said with a demanding tone.

"No," she said firmly.

"I would never disrespect you or the medicine, but I can handle it. Please give me that bottle and let me lie here in my vomit and piss if necessary. I've come too far and can't leave here with these things still inside me." I protested.

"I can't do that, Matt. We are going to eat soon and celebrate the two birthdays." She reminded me.

Beatrice had a son named Matthew, whose birthday was also on the 25th. Later, I would understand that she had shamanically facilitated my rebirth, and I was one of her two sons born on the same day and sharing the same name.

"I want more ayahuasca!" I said unwaveringly.

"What you need is to go for a walk and get yourself under control, and I need to go check on dinner. Join us when you cool off."

I went to my room, packed my stuff, and walked. No goodbye, no asking, I just left. I was again in fight-or-flight and did not want to allow fight back to the surface, so I chose flight. I made my way to the circle to get my remaining items, where I ran into another guest named Rob.

"Are you looking forward to the party tonight?" he asked.

"Nope," I said flatly. "This is all bullshit!"

I finished packing my yoga mat and left my turkey feather and staff beside my spot.

Seven miles to the nearest town was an easy day for a Green Beret. I noted the time and calculated that I'd be there by dinner. My mind was racing a thousand miles a minute. I felt angry, betrayed, and stupid. How had I allowed myself to believe that I was a shaman after three days? Why did everyone else let me believe this nonsense? I needed to distance myself from that entire experience as quickly as possible. I walked for two hours before arriving in town. I spotted a hotel and walked in. I went straight to my room, dropped off my backpack, went back downstairs, and went directly across the street to a diner.

The hostess took me to a table, and I was ready to order before they could bring my water and utensils.

"I'm ready to order," I said to the waitress.

"What will you have?" she asked.

"I want two sirloin steaks medium rare. One with a baked potato, one with fries. One with soup and one with salad." My tone was cold and firm.

The waitress gave me a look, questioning my order. I just smiled back at her. Shortly after, I found myself with a basket of homemade rolls, a bowl of soup, a dinner salad, a baked potato, an order of fries, and two sirloin steaks. I focused on the meat. It had been over a month since I'd eaten meat; my body was protein-deficient, and I sampled some sides. An older couple with a young grandson entered and sat in the booth near me as I ate. The man was wearing a hat that said, "Vietnam Veteran."

Really? I thought to myself. *Really, universe?*

I tried to focus on my meal while ignoring my fellow veteran. I paid my check and rose to leave. Before leaving, I stopped and looked the man in the eye.

"Vietnam vet?" I asked coldly.

"Yes," he replied defensively in response to my tone.

"How do you do it?" I asked through moist eyes.

"How do I do what?" he asked back.

"How do you cope? How do you manage?"

His expression softened. "You just find a way, son," he softened.

His wife stared at me. "Are you okay?" she asked with sincerity.

"No," I said. "No, ma'am, I am not. Sorry to have bothered you." As I spun and walked away to hide the tears I could no longer hold back.

"Get some help, please," I heard from behind.

A gas station was on the corner between my hotel and the diner. I walked in, bought a twelve-pack of beer, and downed two beers as I waited for the light to change. I continued drinking while I showered, then turned the TV on to distract myself. There was a UFC fight. *Perfect*, I thought. Meat, beer, and fighting; it was time to reconnect with the old me. I missed him. Reacquainted with my old vices, I passed out cold.

The plan had been for Sue to pick me up on Sunday morning. I'd refrained from contacting her until that morning. I woke up in my hotel room and sent her a text.

"Hey, change of plans. Don't pick me up where you dropped me off. Pick me up at this hotel instead. I'll explain later. Love you." I texted with a dropped pin.

She texted back, "Okay. How did it go?"

"Too much to try and explain in a text. See you soon." I replied.

I ate breakfast and called John while waiting for Sue. For an hour, I vented to John over the phone. I tried to express how frustrating it had been for me to build up so much hope for this retreat, believing I could cure my PTSD, only to temporarily lose my mind, be humiliated, and end up more traumatized.

He could only say, "Man, I'm jealous, a shaman . . ." I wanted to punch him.

I returned to my hotel shortly before Sue arrived.

I began, "Please understand that what I am about to tell you, I'm still trying to wrap my head around. I probably won't have many answers. But I'll TRY and answer any of your questions."

"Okay." She responded rather nonchalantly.

"But first, is there anything you would like to tell me?" I asked her, double-checking the reality of Kyle's intentions to connect with her.

"As a matter of fact, there is," Sue replied. "I connected with Kyle yesterday."

"Really?" I asked.

"Yeah. I was trying to find books to give to his friends and couldn't find any, so I said, *'Bud, I need to know where to find the books you want me to share.'* Just then, I felt an urge to re-check a shelf I had already checked. I went back to that shelf, and there they were, a whole stack of them," she explained excitedly.

I thought to myself, *Damn it. How long am I going to deny this?*

We sat on the couch in my room for a couple of hours. Through constant tears, I told her everything. I tried to describe every event in detail, the entire process, and everything I had seen and heard over the weekend. I told her about Kyle leading me through the veil and how I had been told that I was a shaman. I explained how I had lost my temper, walked out of the woods the night before, and ended up at that hotel. She listened lovingly, gently, and with no judgment as I spoke, in typical Sue fashion. We drove home, and I unpacked my things before repacking them for our three-day getaway to Sun Valley, Idaho.

9
SEEKING ANSWERS

"Man is a creature who spends his entire life trying to convince himself that his existence is not absurd." - Albert Camus, from The Myth of Sisyphus, 1940.

Over the next three days, I put Sue through constant hell. I was not a fun vacation partner. I was in a non-stop internal wrestle to make sense of what happened. I examined the events of the previous ninety-six hours repeatedly. I continued to reacquaint myself with the carnivorous, beer-drinking, emotion-drowning Matt. We managed to fit in some fun excursions in between ramblings. We kayaked in a cold, deep mountain lake on one of these excursions. We paddled across the lake to the far side, where a small beach was nestled in a grove of pine trees. We walked along the beach, enjoying the scenery. The mountain was covered in lush green pine trees and snowcapped peaks. I looked down into the crystal depths of the lake. The water was clear with a slight blue tint.

"I'm going to go swimming," I announced.

"You didn't bring a swimsuit," Sue replied.

"Not a problem," I said.

Fully naked, I waded out into the glacial water. I felt strong, refreshed, and peaceful. I felt strangely connected to Je, as if he guided me. It had been three days since my last dose of ayahuasca, and I was still perceiving the non-physical reality of an ancestor observing me, waiting patiently for me to accept my identity and stop trying to ignore my mission.

After four days of listening to me nonstop, I asked, "Sue, what do you think?"

In typical Sue fashion, she patted my chest twice and said, "I'd get a second opinion, big guy."

I spun around and walked away. It was good advice. If I were, in fact, shamanic, another shaman would be able to confirm or deny this. I set an appointment with someone I found named Joan for ten days in the future. Joan asked me to journal and focus on my intention during the days preceding our appointment.

The day before meeting with Joan, I wrote the following in my journal: *My intention is to come away with absolute clarity about my purpose and mission in life.*

I had both anxiety and excitement as I drove to the meeting. I wanted answers.

When I pulled up, I was greeted by a sweet lady who appeared to be in her sixties, with long, dark hair with feathers in it. She was adorned in large rings, colorful bracelets, and handmade necklaces. I immediately felt at ease and was welcomed into her space. I noticed the drums, rattles, Native American art, and carvings. She told me that as a mixed-race person, she practiced Norse and Native American traditions. I thought of my own Norse ancestry and how she lived so close to where I'd grown up, and I felt that I had found the perfect person to tell me who I was.

She began the rune casting[1] by drawing nine runes and placing them in order on a hand-embroidered cloth. Then she began to study them by pulling down several reference books and was softly muttering things like "hmmm," "oh, that's interesting.", "wow," and "I didn't expect that." She looked over them. My ego was confirming my suspicions that this was all bullshit.

Then she paused, turned to face me, clasped her hands in her lap, and asked, "Has anyone told you recently that you are a shaman?"

My jaw must have dropped. I came here to disprove what happened two weeks prior, and instead, it was gaining momentum. I told her a little about the events of the retreat.

Next, she began to explain my runes. She started with the difference between my karma line and my dharma line. Karma is experienced through daily actions, the universe's way of showing when one strays from dharma, the destiny path. Dharma is the one thing that I would do if I knew I couldn't fail. She told me that suppressed emotions cause energy blockages and that these blockages would become negative entities.

My center rune related to the physical world. The stone was upside down, symbolizing that my roots were disconnected, and I didn't feel at home. The northern rune related to humanity and showed a connection to the moon. The moon is connected to intuition, which tells us that something is coming concerning serving mankind. The northeastern rune is known as the place of the Gods. It relates to one's highest hopes and dreams or life purpose. The symbol there communicated ice, isolation, and individuality.

The East is the home of giants. The shaman rune lay here. Elk antenna, receiving God's spirit. The southeastern position is the subconscious mind, taken by the Dagaz rune: daylight. It was a

---

1. Runes were are collection of small stone, bone, wooden tablets inset with ancient Norse symbols and alphabet

positive message that the sun will shine on my activities, and I will not fail. In the south was the direction I should take to make things happen. The Raido rune was upside down in the south, meaning I needed to change direction and turn around. In the southwest lived Loki, the keeper of the underworld, half alive, half dead. The shaman read that hidden parts of myself were influencing me, and I was committing self-sabotage. Something wasn't working, and I needed the element of fire to find a solution. In the west lay the rune Ansuz, telling me to communicate: God's mouth. It was important to listen to others and my intuition above all.

Lastly, there were two runes in the northwest position. One related to light elves and karma, a higher ancestral consciousness, and my next path. The second told us that I was moving toward my destiny and experienced deep emotion in the waters of life. Together, this meant that I felt my ancestors' deep, raw emotions as they guided me.

What stood out to me the most was the shaman rune, the connection to the moon in light of my new name, Moon Child, and messages of success. There were hidden parts of myself sabotaging me, and I would need to follow my intuition in a different direction to remain on track with my destiny. As long as I did that, I would not fail my shamanic mission. I looked at the runes communicating ancestral teachers and thought of the long line of ancestors I'd met during my journey and what I had learned from them. I was also particularly drawn to the mention of Loki. Had I fulfilled the intention I'd written in my journal? Was I a shaman? This rune casting seemed to strongly confirm it.

Next, Joan taught me to pray. She asked me to follow this ritual for seven days in a row: I was to begin by facing the east, honoring my ancestors, and thanking them for new beginnings, new seasons, and the new life that they bring to each of us. Next, I would turn to the south and honor the spirit of work that the sacred ancestors represented. I would then turn to the west, honoring my ancestors and expressing gratitude for the spirit of harvest and abundance.

Turning to the north, I would honor the ancestors and the spirit of rest. Lastly, I would point to the sky, connecting with the Father Sky, and then point downward, connecting with the spirit of Mother Earth, and then turn inside, expressing my needs. I agreed to this ritual, and we parted ways.

I wrote in my journal:

As crazy as this all sounds, I am beginning to believe this. I might be a shaman, a healer, a medicine man. I need to change my heart, or rather, allow my heart to change and allow that change to produce better thoughts, words, and deeds. I need to discover and improve my talents and gifts. I need to be confident in what I am doing. I need to make a to-do list. I need humility. I can't share this with just anybody. Hardly anybody. This is a gift to heal, to help. Not a status symbol. Move through the world calmly, confidently, and quietly. Humility every day. Humility and confidence.

Shaman: a spiritual leader outside the norm of religion. The longest journey begins where you stand. Draw power from the web of life; we are all connected.

First Hero's Dose of Psilocybin

Still feeling overwhelmed and confused, I decided to try psilocybin, but this time with a hero's dose. A hero's dose is larger than five grams and lifted from Greek mythology and classical literature, where an exiled person embarks on a perilous and potentially transformative journey into the unknown, confronting their own consciousness, fears, and deeper layers of the psyche, and returns as the victor. I reached out to friends who lived about four hours away and asked them to trip-sit for me.

The night before the journey, Sue woke me up in the middle of the night. She leaned over and said, "I don't know why I am supposed to say this, but 'Peacock.'" Random channeled messages are not something Sue does.

"Okay," I said, still half asleep and confused, and went back to bed.

The following morning, I was driving to the location for the ceremony, contemplating how insane this was. My ego was in full-blown rational mode. I had a thousand reasons why this didn't make sense, why I should just pursue a post-military civilian career and walk away from all of this. My intuition was doing its best to keep up and follow the mysterious chain of events unfolding for the past two months, but my ego was winning. Two hours into the drive, I surrendered and started looking for the nearest exit where I'd stop and grab a cup of coffee, turn around, and go home. I pulled off the interstate, and there, beside the truck stop, was a pen of peacocks. PEACOCKS!

"Seriously, universe?" I said out loud.

I bought a drink and got back on the interstate; two hours until my destination.

We had no idea what we were doing. This was their first time "holding space," and my first time trying to journey on my own. I tried to listen to my intuition and decided on 7g.

Carter and Misty asked, "Are you sure?" well aware that this was a large amount, especially for my second time.

"Yes," I said, "something just feels right about it."

We ground it up with a coffee grinder and soaked it in lemon juice for 20 minutes while discussing how we envisioned the ceremony going. Soon, it was time. I took the cup and placed my hand over it, speaking my intention into it like I had with ayahuasca. My intention was to learn more about my life path; I laid down on the couch.

Within 30 minutes, my head began to tingle, and my belly felt nauseous. I stood up and went outside in the backyard. Before long, I was back in the pure euphoria, a sense of overwhelming bliss and peace accompanied by the pastel colors and mystical geometry. I was happy and surprised to see that psilocybin shared commonality with ayahuasca. It continued until the vision unfolded. I was shocked and somewhat comforted when Je appeared.

He sat before me and said, "I want to show you some things."

I nodded, accepting whatever it was I was going to be shown. I saw Kyle again; this time, he was a young Native American in a war party. The leader of the war party wasn't very capable, and everyone in the war party knew it. The two war parties were on opposite sides of a large valley, but in the center of the valley was a small hill. It was clear that whoever controlled the high ground would control the battle. The enemy began riding out toward the high ground first, but Kyle's leader froze; they would lose if something weren't done. Kyle spurred his horse and rode as fast as he could without thought. Soon, the rest of his party followed suit, spurred on by his actions. Before long, they controlled the high ground and the battle. In an instant, that scene faded away.

Next, I saw Kyle again ride out on the same land, this time as an old man. His hair was long, gray, and braided. He rode a white stallion, carrying a long spear. He was adorned with a massive headdress and a bow strapped to his back. He had become the chief, no doubt due to his actions I witnessed a moment ago. Now, he was tired, afraid of death, and struggled to sleep every night. His demons—the faces of the men he had killed—waited for him just around the corner each night.

"Go to him," Grandfather said.

And so, I went.

"Kyle," I said. "It's me. Let me show you what to do."

With Je instructing me, we cleared the ground. Next, we used our bare hands to dig a grave, and once it was sufficiently deep, we placed his bow and spear at the bottom. Then, stones appeared, and the faces of his enemies were carved on them. We placed the stones on his weapons and worked the dirt into the spaces between. We covered the entire mound with dirt and prayed for Mother Earth to heal those wounds. Once we were finished, Kyle mounted his horse and rode peacefully to the north. I could tell by how he sat taller and more confident that he was no longer afraid. He would be able to cross over and sleep for eternity, and the demons would stay with Mother Earth. He glanced back one last time, and I saw a slight smile on his face before he rode out of sight.

"Good work," Je said, "We have more work to do."

Suddenly, a mountain man from the nineteenth century appeared to me. I saw him with a flintlock rifle, dressed in furs, riding a horse and leading a mule behind him. I saw the terror in his face. I leaned in, searching for more clues. I saw visions of atrocities committed by this man. I saw the blood on his hands.

"Who is this?" I asked Je.

"This is the man who had owned the land you own, but in a previous lifetime. He, too, was afraid of sleep and plagued by demons. He had been afraid to sleep each night of his life and was also afraid to sleep eternally in death. Go to him," Je said, "and dig his grave. Bury his weapons."

I went to the mountain man in the same manner. I met him on the land he owned in his most recent incarnation. He looked at me, dismounted his horse, and we cleared the ground together. We dug the grave, we buried his rifle and knives with the stones bearing faces of the men and women he had killed, we worked the soil into the crevices and covered the stones, and we prayed to Mother Earth. The man got back on his horse, changed direction, and rode north.

Lastly, I was standing face-to-face with myself. The process repeated itself. I saw my own atrocities. I saw forty months of combat. I saw the faces of Iraqis and Afghans, the same faces that would haunt me each night, waiting for me to sleep, wanting revenge. I could clearly recognize some faces of men from those countries, men who wanted to kill me. We—me and the other version of me—followed the same process.

We cleared a space on the ground, dropped to our knees, and used our hands to dig a grave. When it was finished, I took my M4 and my 9mm pistol, the two weapons I'd carried for a lifetime, weapons that felt like an extension of myself whenever I held them. I said goodbye to the weapons and gently placed them at the grave's bottom. I grabbed the boulders engraved with the faces of men who wanted me dead and put them on top of my weapons. As I did so, I noticed that some of the boulders were missing faces—in their place were words like "abandonment," "trauma," "shame," "excommunication," "sexual trauma," and "divorce." I looked up at Je.

"Are these not your enemies, too?" he asked. "Put them in the grave."

And so, I added those stones to the grave with the others, worked dark, black soil into the empty spaces, and buried my demons once and for all.

"There," he said with a sense of closure. "It is finished. You will never see them again unless you dig them up. If you will leave them there, leave them alone, and you will never be haunted by these demons again."

Suddenly, a conscious thought interrupted my channeling. I became acutely aware of the spectacle this had become. I was sitting on the grass in a puddle, pouring water over my arms and head, surrounded by empty bottles. Misty was sitting in front of me, listening attentively as I spoke out loud, channeling Je. I remembered the neighbor I had seen a few hours prior and panicked. What if he took his dog outside again? What would he think of this spectacle?

Grandfather spoke up. "Look," he said, motioning behind me to my left.

I turned to look and saw a wall of ancestors standing shoulder to shoulder, spears in their right hands and shields in their left. There were both men and women dressed like Vikings, adorned in animal skins.

"Who are they?" I asked Je.

"They are your ancestors. They are protecting you. Now, close your eyes, and let's continue," he replied.

Grandfather had placed a shielding wall around me. He had protected this space. He had thought of everything.

I began to go deeper. I could recognize that Je was speaking through me, but at times, I was so lost that it felt as if I were him. I was able to transition in and out at will. Now that Je had taught me how to heal, he transitioned from the healer to the teacher. He had messages for me and messages for others, which he passed along through my voice. He said that Mother Earth was a healer. Like a womb, the earth provided everything we needed: air, water, nutrients, safety, protection, and healing. He said that the earth loved us and wanted to care for us. This would go on for hours.

I felt overwhelmed and lost energy quickly. Carter had set out a couple of small water bottles, and I emptied them over my arms. It wasn't enough; I wanted the icy water inside my veins and needed to feel my Norse ancestors. I dumped water over my body or chugged it until I had gone through several liters. My body didn't feel hot, but the cold energized me. I was impatient and asked to rest, but Grandfather was more impatient.

He said, "I have waited millennia to speak. Your rest can wait."

"Give me five minutes," I pushed back.

I rose, transitioning out of the channeling state, and sat in a patio chair. I hungrily snatched a blueberry muffin from the table and ripped the top off. I managed one bite before Grandfather returned with force and continued speaking.

He said, "Animals are integral as well. They have medicine, just like plants. Understanding and respecting animal medicine, like drums made from hides and sinew, feathers, and consuming animal flesh in season, is important. Using the horns of animals near extinction, as in Chinese medicine, is unacceptable."

He taught me that nature, plants, and animals have the capacity to heal and provide for us as extensions of the earth. More importantly, these entities can heal themselves, given humanity does not interfere.

"We have a responsibility to use nature to bless the lives of indigenous peoples, not to exploit indigenous persons."

Next, Je showed me an image of a book. I leaned forward to focus on it, reading the title: *From Green Beret to Shamanism, One Man's Journey to Heal*. On the cover was a photo of me taken during my last patrol in Afghanistan.

"What is this?" I asked.

"It is one of your missions." He replied.

"Am I supposed to write this?" I asked in shock.

"Yes." He said, continuing his teachings before I had time to soak it in. "Father God in the heavens is the seed, and Mother God is in the earth. Together, they combine to create life. The root of all humanity is singular. Just like a tree has a single seed and many roots, so does humanity. We all came from a single seed, then roots were formed, which spread out to Africa, Asia, Australia, the islands, Europe, South America, and North America."

He quickly showed me several key individuals in my past that I had met and how they had played a role in my life, shaping events that led me to this place. Then he showed me myself. He told me that I was of

Germanic and Norse descent and that I had been born near the summer solstice for a reason.

He showed me my birth mother, Jannice, and all of her pain, agony, and shame. I felt such deep empathy for her. I could feel her stress, worry, and loneliness. I felt her motherly sorrow. For the first time since learning about my adoption, I could feel what it was like to be in her shoes. Up until now, I had only understood intellectually. She had talked with me about each of those feelings and the impossible dilemma she had faced when placing me up for adoption, but never had I understood these things on an emotional level; my ego got in my way.

It felt like every cell in my body knew in an instant the experience of being an ostracized, abandoned, lonely woman facing the most difficult decision of her life, doing her best, and still feeling empty and hopeless. I grieved for her. I loved her unconditionally in an instant. I vowed to call her as soon as I could and apologize for not being more compassionate and understanding, and not having true empathy for her.

Je showed me Mother Bear, my second power animal. He told me to have profound respect for the ferocity of this animal. He said that Jannice had acted as Mother Bear for me, had fought with the ferocity of a bear for my future, had defended me, and had placed herself as a sacrifice in my path when she'd decided to give me away. Mother Bear is a symbol of loyalty, and that loyalty takes many shapes. Jannice had been loyal enough to sacrifice her comfort, wants, and hopes for my benefit.

Next, Je showed me my two other power animals: Grandfather Moose and the snake. He taught me that the moose is steadfast and strong. The moose is born to work and has the astounding ability to move forward each day—old, slow, wise, strong, and resilient, day in and day out. My path was filled with work, from the grueling days of military training to mystifying spiritual study and shadow work.

Grandfather Moose is my model of persistence, inspiring me to labor each day for my growth and the growth of others. My third spirit animal, Snake, was feminine, so it was given to me. I was beginning to encounter the divine feminine and would need to balance my masculine and feminine energy. Snake was here to show me that balance and to teach me how to shed my old skin and grow anew.

Je then showed me my birth father. Ted's role was to be unavailable. It goes without saying, but he fulfilled it brilliantly. I don't mean that sarcastically; sure, it was painful to be abandoned, and his actions contributed to my problems, but there was no path I could have taken to arrive where I am now other than the path Ted started me on. That was his role. I will honor and respect that, but I will also balance it with what Je taught me next. There is great devastation being experienced by the vast majority of families all over the world due to abandonment issues.

He said, "We cannot change the world, but we can change ourselves, which in turn changes the world."

Je brought back my power animal, the moose. Moose, the steadfast and strong, like granite, ever marching along. "This is how a father must be. Stalwart, hard-working, protective."

I looked up at my hosts, Carter and Misty, who had been so loving and patient during this journey. Running back and forth, retrieving water from inside the house, and bringing it to me, where I sat on the lawn. I felt immense gratitude for them both.

I dove back in, suddenly feeling the anxiety I had early that morning and the identity crisis that had been ongoing since my first journey with ayahuasca. Je sensed what was on my mind.

"Ah," he said, "you are wrestling with your calling as a shaman, aren't you?"

"Yes," I answered.

"Trust yourself," he replied. "Everything you need to know can be found within your intuition."

"How can I tell what my intuition is?" I asked.

"Your head thinks, and your heart feels, but your gut KNOWS," he said. "That is intuition."

"Okay," I nodded.

"When all three confirm the same thing, it is absolute. The more you use it, the stronger it will become."

Internally, I began to question him. Everything in my life was related to logic and reason, and my training was in making sound decisions. I'd never been allowed to say, "I think we should patrol this village because my intuition says so." No! I had to have sound reasoning, facts, and explainable logic.

"No?!" Je shouted, interrupting my thoughts. "'Reason kills intuition!" he roared.

Despite possessing wisdom and knowledge gathered over several millennia, Grandfather was still short-tempered—especially when I focused on myself or questioned him.

"When you say 'no' to intuition, it stops working," he explained. "Not only does it stop intuition, but it also stops all your spirit guides. Me, Mother Bear, Snake, Kyle, and others are all sovereign beings. We have free will. If you say 'no' to us, neglect us, or misuse us . . . we will leave you. Learn to trust us, which are extensions of your intuition."

He switched topics abruptly. "Plant medicine is a contract," Grandfather told me, "between the Gods, the plants, the participant, and the shaman."

I decided to ask him a practical question. "How much do I dose?" I asked. "When do I dose? How often?"

"You don't have to know," he said. "Your intuition will know! Do you remember what I just taught you about intuition? I'll remind you. Plant medicine is not to be trivialized, not to be exploited, not to be abused. It is to be used in moderation."

He taught me how every culture on earth has access to DMT.[2] It was found in animals and plants on six continents. He said that keeping hallucinogenic medicine from the people was evil. It was evil in the form of greed, politicians, and Big Pharma. Plants were intended for the use of man, and to restrict them was pure evil. This extends beyond plant medicine—all plants.

"You cannot use plant medicine for recreation," he opined. "Plant medicine is sacred and should always be treated as such. It should be used sparingly, only as needed, never for recreation."

Then he asked me, "Why do you want to use DMT next week?"

I had been planning on trying smoking DMT with Gary, who had been a co-facilitator at my ayahuasca ceremony. I ran through my intentions with him.

"I want to experience all the medicines I can if I'm meant to be a pharmacist of sorts. I need to know their limitations, side effects, and purposes," I explained.

"Those are good reasons," he said. Sensing I had more, he said, "Continue."

"Secondly, I want to meet God," I said. "I am agnostic, and I need to know if there is a God."

"Really?" he snorted. "Do you really think that all of this is

---

2. There is a more detailed explanation of DMT later in the book, but when DMT is smoked it is the most powerful dissociative psychedelic in the world. It is used sparingly.

happening without a God? God is love. Do you feel loved when you take medicine?"

"Yes," I answered without hesitation.

"There," he said. "There is God." He gentled his tone. "God is pure love. God is warmth. God is eternal." He paused for a moment before continuing. "There are many Gods. Father God is in the heavens, and his partner, Mother God, is in the earth. Can you not see the perfect partnership that is formed? Do you not see it?" he asked.

"As above, so below." The phrase came to my mind suddenly. I had heard it many times during my ayahuasca journey.

"Correct," he said. "It was a perfect partnership, neither one superior. Each fails without the other. Even with Gods, it takes both."

My mind flashed back to Tom, and I wondered how, without his contribution, I would not exist as I am.

"That's right," he said, interrupting my thoughts. "It is all connected. Do you see? Do you see now why a father and mother are equal partners? Neither is superior to the other. They are equals. Both are needed."

My mind wandered. "What about religion?" I asked.

"All things require balance, my son. Balance and centeredness are key," Grandfather said.

"What do you mean?"

"Religion has its place," he answered. "In many cases, religion provides moral teachings and social structure. It has even preserved critical teachings and books."

I nodded, signaling that I understood.

"But extremism in anything is destructive," he continued. "Extremism in religion was responsible for the rape of the land, the murder of Indigenous peoples was fueled by greed and lust, and caused sexual repression. Everything in life comes down to balance and centeredness," he repeated.

His voice became stern as he emphasized, "Everything. Everything in life will seek balance and centeredness. It is the way of nature. All religions are man-made, and how they interrupt spirituality. In the absence of spirituality, man-made religions fill in. The intention may have been pure, but they can all go astray and become extreme. Ultimately, religion became an enemy of spirituality. Religions killed the shamans who understood the universe, the healers with their plant medicines, and those who could speak with animals. All were killed because of religious extremism and the need for control. The leaders of those religions who were jealous of miracles, spirituality, power animals, and plant medicines eradicated them all. Find the center. Find the balance. Avoid the extremes," Grandfather concluded.

Next, he taught me about responsibility and respect. "Your societies have it all wrong," he said. "You're focusing on the grandiose side of the problems, and you'll never solve them or change anything that way."

"How do we solve our problems?" I asked.

"It's about leverage," he answered. "You want to fix climate change, or the economy, or tyranny, or generational trauma by trying to change the world. You have neither the leverage nor the fulcrum to move a mountain that large. Focus on yourself. Focus on fixing yourself. By taking responsibility for your own actions, you begin to change yourself. You can only control yourself."

He switched topics. "Respect," he said, "is more important than love. Stop seeking approval and validation—stop worrying about likes, what's trending, and viral stories. Do what is respected and do what is right. Seek the respect of others—not validation."

My body was stiff and tired from channeling for six straight hours. I had lost track of time but knew I had been sitting too long for my old, arthritic bones to handle. I stood up and stretched, then walked in a small circle in the yard, trying to comprehend everything I had heard and seen, the process of channeling, and how it worked. I ate another muffin and some fruit and sipped some water. Grandfather was more patient this time; he could tell I needed a rest. But he was only slightly more patient.

He broke in again. "Before we end, I want to give Carter and Misty a gift."

I made eye contact with Misty and said, "Misty, Grandfather wants to speak to you now."

"Okay," she responded hesitantly, unsure of what might exit my mouth considering all that she had seen in the course of the day.

Je used my voice when he addressed her. "Misty, I see you. I see you in a past life. Your name was Madeleine. You were a seventeenth-century French medicine woman, a gypsy. You had a wagon that you lived in and sold medicines from. Two large horses pulled it. You were loved. You were loved by all the villagers you visited. You would travel from town to town, selling plant medicines and telling stories. Everyone waited all year long for you to return."

"Ahh," she responded, a small tear in the corner of her eye.

"And Carter," Je continued, "You were Danish in a past life," he began. "You were the younger cousin, and your older cousin was the elder of your clan. But he was a poor chief. No one liked him, and he couldn't lead. You knew your place, and although you were naturally better than your cousin, you knew that you couldn't be king and had humility. You followed behind your cousin, keeping everything running smoothly."

"I see it too, Carter," I said. "I see how much you have done for me today, and I deeply respect you for allowing me, a practical stranger, to be in your home. Thank you."

AYAHUASCA RETREAT NUMBER TWO

I was beginning to think this shaman thing might be real, but I still had countless questions. Beatrice was still not speaking to me. Who could blame her after the way I'd left her retreat? I decided to try to find another ayahuasca retreat. Once I discovered that ayahuasca retreats existed in the US, I found many. I chose one and booked my flight.

On the way to the retreat, I stopped at a national park, the birthplace of Abraham Lincoln. I thought of how much of a hero Abraham Lincoln had always been to me and how he was a symbol of never quitting. I thought a lot about slavery and emancipation. I thought about how the mental slavery of PTSD impacted so many of my brothers and sisters. I thought of the emancipation and freedom that ayahuasca offered—freedom from torment, addiction, and pain.

I felt spirit coursing through my body as I strolled the beautiful grounds, taking in the sight of ancient hardwood trees. I tried to speak to the trees. I wanted to feel their energy and tap into their old wisdom before the retreat. I spent a considerable amount of time there before deciding to visit a nearby state park to meditate and make use of the remaining time.

The state park had a beautiful lake with large granite shores that were as if they had been carved into a quarry. I walked north toward the shore, sat on the grass, crossed my legs, and got comfortable. I noticed a dragonfly hovering around me. It was another synchronicity. The dragonfly held meaning for us, reminding us of Kyle. I looked directly overhead and spotted a flock of crows circling above me.

I had the sense that a spiritual blessing had been placed on me. I began to clear my mind, working through each chakra in order, from bottom to top. I recently started a meditation practice focused on my breathing and the color associated with each chakra. This practice allowed me to clear my mind, align, clear negative energy, and balance myself. The purpose of this meditation was to focus my intention for the ceremony. I had set the intention of gaining clarity about this 'Shaman' thing.

I noticed a strange feeling as I closed my eyes and began my breathwork. It felt familiar but also like a distant memory. Suddenly, I began to float out of my body. I could sense my spirit, but simultaneously, it looked down on my body. In a split second, I was jolted into a memory of having out-of-body experiences as a child. I remembered asking my peers and the adults in my life whether they could do it too, and being either quickly dismissed or mocked. Ultimately, it was my father who, despite being well-intentioned, told me that it was satanic and never to do it again. And with that, I stopped having out-of-body experiences. But, again, it was as if a long-forgotten ability was returning to me.

The feeling lasted around ten minutes before I returned to my body. Instantly, my mind was flooded with a thousand deeply repressed childhood memories. I hadn't been aware of them since that conversation with my dad, whose approval I sought so badly—forty-year-old memories of leaving my body, talking with trees and animals, and seeing spirits. In what felt like a split second, I fully recalled all the gifts I once had. For the first time, this "shamanism" idea was coming from an internal source, not an external source. I gently opened my eyes to see that more crows had joined the flock overhead and that more dragonflies and butterflies were floating above me. I laid back and stared into the sky, watching them all silhouetted against a deep blue sky and white, billowing clouds.

As I drove silently the last ten miles to the retreat, a cardinal fell from the sky and landed on the road directly in front of me, but by the time I pulled over, the cardinal had disappeared. It's unclear whether this was an omen, a blessing, or a random event, but the bird was gone, and I continued on my way.

I arrived at a steel building on the side of a state highway and pulled into the parking lot. The building could have passed as a daycare or dance studio, but it was neither. A young lady met me in the parking lot and pointed a thermometer at my forehead to check for a fever. I grabbed my bags and followed her inside. There were two main rooms: a small room used as a classroom and the ceremony room. The ceremony room had small, square mats covering the floor and halfway up the walls. Four partial walls provided a degree of privacy, like cubicles. The cubicle-like sleeping areas contained antigravity chairs and fold-up beds. Small red buckets accompanied each chair. The walls and ceiling were black and decorated with various hand-painted lotuses and mushrooms. Many psychedelic-themed tapestries hung around the room. I was given the cubicle on the far side of the room, directly facing a large tapestry with a large skull and an owl. The adjacent classroom was set up with folding tables in a horseshoe formation. Three women were seated at the tables.

The next couple of hours consisted of group introductions, instructions about ayahuasca, what to expect on the journey, and discussions about dosing. One of the facilitators, Shelly, instructed us that we would receive a small dose, amounting to about two tablespoons, to begin. After we had taken that dose, they held us in the classroom for thirty minutes to see how we reacted, then administered a second dose of the same size. Combined, the doses equaled about one-fourth of the dose I had taken the first time.

They guided us to the larger room after the second dose. I sat on the floor in my space and used my newly acquired rattle to drum quietly to myself, trying not to make a scene or distract the others. I felt the familiar feelings return as the medicine began to flow through my bloodstream. The familiar visuals returned: the fractals, geometry, and neon colors. But that was all. I felt disappointed. I went to the bathroom, and when I returned, Shelly confronted me. She instructed me to stop drumming and sit in the chair. I learned quickly that not everyone who called themselves a shaman was worthy of the title and that each ayahuasca experience differed greatly. I returned to my chair, hoping that something—anything—might happen, though I had little hope for the minuscule dose I had taken.

A moment later, I began to feel energy build within me and fill my entire body from head to toe. Then I saw Her. Ayahuasca is referred to as "She," "Her," "Mother," and "Grandmother" interchangeably. I had failed to see "Her" on my first two journeys, but believed what others had told me, that ayahuasca is a feminine entity. Now, I was face-to-face with Her.

I heard the most beautiful South American music from somewhere deep in my mind. I listened to a pan flute and drums as I kept my eyes on Aya. As the volume of the music increased, I began to dance in my chair, my hips swaying back and forth, and my feet moving to the Latin rhythm. I danced as if I had been dancing to this music all my life. I suddenly had a conscious, panicked thought: *What if they see me dancing, having been warned once already?* To my surprise, as I opened my eyes, I saw that my body was lying perfectly still. I had left my body for the second time that day. The medicine wore off shortly after, leaving me with a slight sense of euphoria. Even with that small dose, spiritual gifts were returning to me. I saw the spirit animals and the past lives of both facilitators and participants.

The next morning, I grabbed my journal to write about the previous night's experiences. After journaling, I went outside to pray and meditate. Later, Shelly reviewed the procedures and reminded us how to handle the journey through breathing and focusing on our intentions. She added a new precaution.

"If you feel yourself astral projecting, don't panic," she said. "It will feel like you can't breathe or your windpipe has been constricted. It hasn't, I promise. It just feels like it. Just force yourself to breathe. Just take a breath."

Shelly administered the medicine before we returned to the larger room to sit in our chairs. I relaxed in the chair, enjoying the calm, gentle euphoria as it grew in my body. I asked the facilitator how long it had been, and she informed me that about thirty minutes had passed. I thanked her, noting the distortion in my perception of time. Then, almost immediately, I felt it.

Here She comes, I thought to myself.

I opened my eyes and stared at the skull on the tapestry in front of me. Then, a giant snake appeared and slithered up the wall. It maneuvered in and out of the skull and rested coiled on top. I remembered that Mother Aya was also described as a snake, and there She was, serpentine and beautiful. Next, she morphed into her human form, portraying a Latina woman with long, thick black hair, beautiful, piercing eyes, and dark skin.

"Dance with me," she said.

"Okay," I replied.

I rose from my chair and began to dance in the darkness. I had suspended my concerns about the others seeing me dance. I danced as though I were Antonio Banderas. It was a celebratory dance. As I circled and twirled around the room, I noticed my body lying perfectly still in the chair, covered in a blanket, with a smile on my face.

As that realization settled in, She laughed at me. "You're doing well, my son," she said.

All I could do was smile back at her.

"Are you ready for what's next?" she asked.

I nodded, watching her intently.

"Okay. Next, we'll astral travel."

"Astral travel?" I thought.

Instantly, I recalled how dismissive I had been of the warning only hours before that astral travel might happen; there was no turning back. I relaxed and allowed the medicine to take over. My throat began to constrict. It took every ounce of concentration to focus my thoughts on breathing. I could feel the rushing wind in my face, making it more difficult to get air into my lungs. It was like sticking my head out of the window of a jet and trying to breathe deeply. My breaths became very labored and deliberate. I struggled and gasped as I felt myself leaving my body. At some point, breathing became natural again, and I found myself traveling to the center of the universe. We passed planets, stars, and the Milky Way. By directing my mind, I could journey anywhere I wanted.

Grandmother Aya returned to me and asked, "Are you ready for what's next?"

"Yes," I said softly, trying to imagine what else she could have in store.

Grandmother whispered, "Open your eyes."

I did, and I could see through the ceiling of our building. I was gazing into the night sky, looking at the galaxy. Instead of seeing the galaxy as I know it, I saw a digital version with a matrix superimposed over it. There were lines forming grids, which connected at "nodes" or intersections. I saw the real universe, not a cosmic improbability, but a masterful design with alien life forms inhabiting planets and the great connectivity of the invisible grid system. I saw life beyond our beautiful world. I could feel the weight of that knowledge enter my mind.

Just then, Pam returned.

She knelt beside me and asked, "How are you doing?"

"Really well," I replied.

"Great. Tammy wanted to know if you would like another half dose." She continued.

"Yes," I said with a smile.

I followed her out to the main room. Tammy poured a half dose.

"Actually, may I have a full dose, please?" I asked.

Tammy looked me in the eye, examining my state of mind, and then poured a full second dose. I took it into my hands while I set the intention: *Allow Grandmother Aya to teach me.* I drank the medicine.

I returned to my corner and settled back into my chair. No sooner had I settled in than I began to hear one of the guests vomiting uncontrollably. My heart went out to him. My mind recalled my first ayahuasca ceremony and how the shaman had allowed me to help. I knew this wasn't allowed at this retreat; these facilitators were much stricter. Then it dawned on me: I didn't need their permission to travel astrally to him. I closed my eyes, settled into my chair, and left my body.

In the Astral Realm, I asked him, "May I connect to your energy?"

"Yes," he replied.

When I connected, I could see and feel the depth of his pain. The word "duality" came to mind. I could see the perfect balance and duality of his mixed heritage. I could see his masculine and his feminine. I could see colonialism and Indigenous ancestry. I could see his struggle to balance all of it. I placed my hand on his heart chakra and saw him as a female arctic snow fox. I held him and saw him giving birth to himself. Then, I saw a magnificent rainbow warrior, a tribute to the rainbows of Hawaii.

"Rainbow warrior, warrior for love," I said.

The vomiting ceased, the energy changed, and he rested quietly.

"Are you okay now?" I asked.

"Yes," he replied softly, and I returned to my body resting in my chair.

Next, I overheard another guest who was struggling. He was releasing excruciating moans. I again left my body and walked across the room to him.

"Hello, Mark," I said.

"Hello," he whispered.

"Can I connect to your energy field?" I asked.

"Yes," he replied.

As soon as he gave permission, I saw him as a young boy, standing alone before me.

"What do you need, Mark?" I asked.

"I need you to go find my father. Where's my father?" he begged.

With that, I left the room and returned to the expanse of the universe. I had no idea where I was going or how to find his father.

I cried out to the universe and my spirit guides, "Please help me find Mark's father!"

Moments later, I was standing back in Mark's cubicle. Mark was to my right, and a grown man was to my left. They were facing each other.

"Mark," I said, "what have you been wanting to say?"

Mark looked at me, then at his father. I encouraged him to speak freely. He glanced back at his father and began to speak.

"Why? Why did you drink?" He started quietly, then his voice began to crack, and he yelled and screamed. "Why did you give me alcohol?"

His father almost immediately began to cry. He didn't resist or get defensive. He knelt and held his son, wrapping his arms around him as he sobbed.

"I am sorry, I am so sorry, son. I didn't know. I didn't know what would happen. I did my best. I am so, so sorry, son."

The two held each other as they cried. Thirty years of anger, frustration, resentment, and hatred melted into love, compassion, and understanding as father and son embraced on the spiritual plane.

"I am going to leave you now," I intruded, and I returned to my body once more.

As I began to settle, I became aware of the final guest who was struggling. I, again, left my body, traveling to her and asking permission to connect with her and her energy. She consented, and immediately, I saw a damaged crown chakra. She doubted her self-worth and wrestled with guilt and shame. I began with her root chakra and worked my way up from the root to the third eye chakra, gently cleaning and balancing each. When I arrived at her crown chakra, I felt impressed to place a "crown of infinite worth" on her. After ensuring I had helped, I returned to my body.

I was hoping to rest when I suddenly felt that Sue was in need. Sue had signed up for a medium training event in Park City, UT. I knew that to help Sue, I would need to find Kyle. Once again, I left my body and went into the universe.

Soon, I was standing face-to-face with Kyle. "Kyle, your mother needs us. Will you come with me?"

He agreed with a large smile. Together, we traveled to Sue. The three of us sat in a circle in a field of wildflowers, enjoying the sacred space. Kyle was about to open the veil for his mother, the way he had been my guide, and I knew I was no longer needed. I asked them if I could leave. Shortly after returning to my body, I was shown Sue parting the veil and looking beyond. I saw her dressed in white with magnificent white angel wings.

I lay in my cot. Covered in my blanket, I had several revelations. I felt Grandmother adjusting my chakras. I was in pain as She worked on me. I began to scream before reminding myself to surrender, relax, and lean into the pain. As I surrendered, the pain subsided, and I felt at ease.

I sat reflecting on my daughters. Next, my thoughts turned to the massive energy surrounding this event. There was a pattern of cosmic energy beginning to form surrounding ceremonies. This event took place on Friday, August 13th. The day before, lightning had struck the building. There had been a massive thunderstorm overhead on the second night and a meteor shower, sitting over the world's largest cave system and directly on top of a time zone line, highlighting the irrelevance of time.

It was over. I had full peace and serenity in my mind and body. I was exhausted. I decided to go outside to seek food and water. A certain reverence filled the space. Tammy, Meghan, and Pam were whispering to each other. I approached and spoke with them. I thanked Tammy for the extra dose of medicine. I thanked Pam for the smudging and bringing spirituality to the ceremony.

"Are you Cherokee?" I asked.

"Yes," replied Pam. "How did you know that?"

Explaining my recent ability to see people, I turned to Meghan and said, "I saw you as one of my Irish ancestors, a grandmother. Thank you for being here and bringing my ancestors to the ceremony."

I began to relate the experiences of the evening to the group. As I was detailing the highlights of my trip, black words on a white background appeared. It was the same as when I had been told "go learn" on the first night, but this time, they read "you passed."

"I passed?" I questioned.

The words reappeared. "You passed."

I was puzzled.

"Yes! YOU PASSED," flashed for the third time.

A lightbulb went off, and I realized what this meant. The medicine was speaking to me in my language. It slowly began to dawn on me that just as I had passed an SFAS-like event for shamans. I realized that I had passed my shamanic selection.

I had been wearing a stocking cap that I now realized was missing. I grabbed the blanket lying wadded into a ball at my side and dug through it. I felt the stocking cap nestled in the blanket folds, and as I freed it, for a few seconds I saw and was holding a green beret. In an instant, it changed back into the stocking cap.

For a moment, everything was crystal clear. The medicine—Grandmother Aya—had put me through my paces. She had shown and helped me leave my body before the ceremony as a precursor for what was about to come. Then, she showed me how to travel among the stars in the grid system of the universe, and she gave me four challenges or tasks to complete independently, just like my experience with SFAS. I was flooded with memories of my actual SFAS. I recalled the words, "Selection never ends." I realized this spiritual journey was only the beginning of an entirely new path, and I would still need to undergo a significant training process.

At times, I have been criticized for having too much military presence in how I conduct myself as a shaman. Still, this experience reinforced a lesson that would play out several times over my journey—my whole life had prepared me for my role. Even the motto of the Green Berets—De Oppresso Liber, to free the oppressed—was still valid. I began to see that this work was to free the oppressed from mental and physical illness. Source would reaffirm dozens of times throughout this process that my Special Forces experience was part of the shamanic training grounds.

Suddenly, as if blinders had been removed, I saw my surroundings for the first time. Everything in that building meant something to me. There was a US flag, an Army flag, skulls, Cambodian statues, Viking statues, African masks, and remnants of Appalachia. I had seen these things without truly seeing their significance. Almost every object in that building had some familiarity: reminders of my career in the Special Forces, clues to my ancestry, and my past. I had gone there for clues, and I found them. Even the steel building was reminiscent of many buildings I had spent countless hours in during my career. There were breadcrumbs everywhere!

I had permission to go outside so that I could continue to walk freely. As I walked, I turned my phone on. I had only one notification from Sue. It read:

Can you believe that they made us read someone on day one of the training? This is crazy, but guess what? I did it. The person I read said that I got everything right.

I texted back:

I know, baby, I know.

10

APPRENTICESHIP

"The master has failed more times than the beginner has even tried."

— Stephen McCranie

The first thing I did when I returned was call Joan and set another appointment.

When I arrived, Joan welcomed me back into her space, asking, "So what did you want to address?"

I explained the events of the last couple of months and the meditation that brought back my memories.

Stammering to find the right words, I asked, "Um, I don't know how to ask this; I didn't see any mention of it on your website, but, um, is there such a thing as an apprenticeship in shamanism, and would you accept me as an apprentice?"

Joan thought for a moment and sat back, folding her arms. For the next hour, she talked about her apprenticeship, the homework, and the structure she followed with her mentor. She also spent a considerable amount of time warning me. She stated that this is not an academic process but a relationship with the universe. She continued by saying that if I accepted this apprenticeship, I was submitting myself to the universe and that it would subject me to the refiner's fire of sorts, bringing all my "stuff" to the surface to be addressed.

She concluded by saying, "It will be the hardest thing you ever do and will call for your death."

There it was again, this warning of death. She said not to take the decision lightly and gave me homework to meditate on my decision for two weeks before contacting her. Immediately, I thought of all the units and training I had attended and how difficult they had been, but something inside me confirmed that she was speaking the truth and that this would be harder.

Joan had been raised in Alaska and had similar experiences with nature, but the benefit of not having to repress the memory of them. One day, when she was eight years old, she wandered alone down a path at a rest stop when she came face-to-face with a black bear. She said that "they looked each other in the eye with mutual respect and that she didn't feel any fear whatsoever." Thus began her journey with unexplainable events and the non-physical reality.

Joan ultimately became a physician and was working in a clinic when an Inuit woman came in and said, "Why are you here?"

Confused, Joan said, "What do you mean? I'm your doctor."

The woman replied, "No, you're a shaman; why are you wasting your gifts with Western medicine?"

Shortly after, I decided to start an apprenticeship with Joan. Our pattern was that I would report my homework to her, discuss it, and then she would teach the next principle before assigning more homework. We continued in this pattern for almost two years. Joan was perfect for me as a mentor. Her petite frame and added age couldn't disguise her power. She could look into my soul and cut through my bullshit with a glance. I was studying at the feet of an absolute master, and I had the utmost respect for her and her wisdom. She was also stern, which resonated with my military mindset; she had standards and enforced them. One of her standards was "no plant medicine." Joan had never once tried any psychedelics.

This rule created a slight dilemma for me. I knew that ayahuasca had saved my life and could reach those who normally couldn't be reached. I felt called to work with plant medicines and knew this would require another apprenticeship. Fortunately, I contacted another shaman who was, by all accounts, the most elusive person I had ever encountered, and that's saying a lot from a Green Beret.

Gabrial was a bodybuilder and pro mountain biker with seven master's degrees from Ivy League schools, all of which were efforts to win his father's love. More importantly, he had apprenticed in Peru, serving ayahuasca for seven years. I sent a text and asked if we could do a phone call. He agreed, and we were on the phone within a couple of days. He started by lowering my expectations.

"As a favor to our mutual friend, I'm happy to listen to you and why you think you are a shaman. But I seriously doubt that you are." He opened.

"Uhm, okay," I replied.

Just as I was starting to believe this was real, I met the first person to cast doubt on it.

"So, tell me what happened," he said, turning the conversation over to me.

I proceeded to tell him about my PTSD, addiction, alcoholism, and arrest, and how I had learned about ayahuasca and was looking for a ceremony.

"Yep, yep, ah huh." He interjected to speed things along.

Speeding up, I told him about night one, the vision of me holding a staff, the turkeys, the turkey feather, and the staff at the bottom of the cliff.

He interrupted with, "Okay. I've heard enough."

I assumed he had just rejected me and replied with "okay," feeling confused.

"Yep, that's enough, you're legit." He replied.

"How do you know?" I responded with more confusion.

"Because your tools have started to find you. In many traditions, the shamans use a staff as a sign or source of their power; yours clearly found you on day one." He explained. Thankfully, he was the real deal and knew a great deal about shamanism.

"Huh," I replied, trying to let that soak in.

"I'll set up a time when we can meet. Talk to you later." He said as he ended the call.

Weeks would go by before I'd hear back from him.

One day, I received a text from Gabrial stating: *Let's meet Wednesday at noon at the Chinese buffet on State.*

Gabrial was very stocky and muscular, had platinum blonde hair, eyeliner, a nose ring, and wore black leather boots with cut-off sweatpants and a pastel tank top—not what I expected. We asked for a table and filled our plates. For almost 55 minutes, we spoke about everything but shamanism.

"Can I hold your wrist?" Gabrial asked, shifting the conversation.

"Sure," I replied, not anticipating what would happen next.

Gabrial closed his eyes and began to channel and loudly speak "light language." Light language is non-recognizable, with tones reflecting energy and subconscious thought more than recognizable words or dialects. The construction workers around us began staring, and I wanted to crawl under the table. My journey into shamanism seemed to get stranger the deeper I went.

"Okay," he said, finishing and sliding the check to me, "I'll be in touch." And he disappeared.

Six weeks passed, and I hadn't heard from him, so I texted our mutual contact, who had introduced us.

Has Gabrial said anything? I asked

No, he responded.

Okay, well, he said he would be in touch, and I haven't heard from him, giving away my impatience.

He has your number!

Okay, I said.

Another six weeks passed, so I contacted our mutual contact again.

I still haven't heard from Gabrial . . . Trying to reinvigorate the conversation.

HE HAS YOUR NUMBER!!! He replied.

I didn't respond.

Several more weeks passed when my phone buzzed, and the text read: *Wednesday at noon at the Chinese buffet on State again?*

I replied with an enthusiastic *Yes!*

Again, we repeated the same process. Fifty-five minutes of chit-chat followed by holding my wrist and more channeling. I was more prepared this time.

This time, Gabrial slid the check to me and said, "Follow me to my car."

Once outside, he said, "Last time we met, I was checking with my guides about whether to work with you. This time, I was just confirming."

I nodded, following along.

"Here, take these," he said, handing me two aspirin bottles marked with a sharpie with the numbers 1 and 2 and handwritten instructions on a napkin from the restaurant explaining how to mix them separately and take them.

"Follow these instructions and get back to me," he said as he spun and disappeared.

Soon, I was apprenticing simultaneously with both Gabrial and Joan. Joan taught me the ancient shamanic energy work traditions without using plant medicine. At the same time, Gabriel taught me about the plant medicine ceremony. They complemented each other very well.

The core principles of the universe were almost perfectly aligned. I can only think of one principle that they had differing opinions on. They were both stern and demanding and held me to standards while I moved through the lessons and homework. I appreciated being able to go to both for advice or to explore the meaning behind my experiences.

In 2022, I was fortunate to meet a Titia. Titia is a title in the Colombian shamanic tradition. A title that meant more than Shaman, it also held a similar meaning as Tribal Elder in Afghanistan. Titia was a fifth-generation shaman who lived on ancestral land near the headwaters of the Amazon River. His "retreat" was raw. The westernization of ayahuasca has created an artificial standard with an expectation of having a masseuse, chef, and yoga instructor on staff. Titia was off-grid in every sense. We were fed from his flocks of chickens that ran around the outside of his hacienda. The ceremonies were conducted deep in the jungle in the Maloca. When we purged, we were instructed to roll over and lean outside the Maloca and keep it in the drainage ditch that circled the structure. What I was able to participate in was identical to how it was practiced thousands of years ago. My experiences were indescribable.

In 2023, I returned, bringing a group of other veterans to participate in the medicine for the first time. While I was there, I asked him to be my third mentor. I've never served ayahuasca and never plan to. I feel that it is not culturally appropriate, but I did recognize that I was in the presence of a mystic who offered me access to wisdom and experience different from what I had learned.

I share the common belief that shamanism is grounded in mentorship. There are several red flags when deciding on working with a shaman; one of the flags is a shaman who has learned everything from YouTube.

Joan taught me that "If you feel like you have learned everything you can about shamanism, you are a danger to the community—get out!"

All my mentors continue to learn from others. When I "graduated" from my studies and Joan honored me with the title "Master Shaman," I politely reminded her that I would always be her apprentice.

In one of my visions in Colombia, I saw a blacksmith forging steel in a fiery furnace. The first thing he forged was a sword, and when he finished, he knelt and presented it to "the universe."

Grandmother whispered, "The sword is the shaman, and the furnace is the apprenticeship process."

Next, the blacksmith returned to his furnace and forged a chain, where every link was both apprentice and master. The more mentors we have, the better.

I continued to study, learn, and perfect my craft. I regularly performed plant medicine ceremonies, conducted one-on-one healing sessions, and did rune castings and spirit animal retrievals. At one point, we held multiple ceremonies each month, averaging up to 12 people per ceremony. During that time, we were witnesses to numerous miracles. Alcoholics, addicts, PTSD sufferers, those with depression, people seeking connection with passed loved ones, or those simply seeking guidance, by and large, were granted their intention. I was performing the multiple techniques I had been taught, including cord cuttings, spirit animal retrievals, chakra alignments, removing blockages, removal of attachments or entities, soul part retrievals, and soul retrievals.

By 2024, I was burned out. I had made mistakes; I had caused trauma in the space where I was there to release the trauma. I was questioning my decision again, and, in some ways, I was feeling mentally ill again, feeling like I did before my first ceremony. Shamanism had become a burden and a curse. Multiple colleagues told me to slow down and take breaks, but I was conditioned to just "muscle through" anything. I saw slowing down as quitting or failure, and that was never part of my vocabulary.

In the space of 96 hours, several signs came to me that convinced me that I needed to listen. The first was a psilocybin journey with a friend that went horribly wrong. During this ceremony, I wasn't letting go; I was fighting it, and it was a miserable experience. Two days later, I was supposed to be at a meeting at 6 p.m., but I had the wrong time; it was at 7 p.m.. So, I had an hour to kill. It occurred to me that I had a friend who did psychic readings at a metaphysical shop not far from where this meeting was to take place, so I drove over.

As soon as we began, she said, "Your guides, angels, and ancestors all say the same thing— 'we are amused at how stubborn you are at doing things your way; let us know when you want to be our partner.'"

I sat back as she continued to channel messages for me directly from the team of guides I had been working with for years, but had

197

ignored. Everything she said was on point, and I knew I had been going about everything my way and was failing. I began to consider the suggestion that I was given—a sabbatical. I decided that I would start doing everything the opposite way. If what I was doing currently had led to this dead end, the way out was to do things in the opposite fashion.

The next day, I went to be interviewed by a woman who was writing her thesis on the modern underground world of plant medicine shamanism.

When I walked in, she said, "Oh wow, your heart chakra is not well; I don't think we should do the interview."

"Um, okay," I said with a degree of frustration, as I was still stubbornly trying to simply get stuff on my to-do list completed.

"I'd like to serve you." She said.

All the red flags were going off in my mind. Things like *yeah, but I am the one who serves others, I don't want to appear weak, I am okay, I don't need help* . . . but then I recalled the idea from the night before —do everything opposite.

I replied, "Sure, what do you have in mind?" assuming she would suggest a meditation, Reiki, or maybe a sound bath.

"I am thinking DMT." She said matter-of-factly.

Shit, I didn't see that coming when I woke up this morning.

"Sure," I said, holding in my fear.

Her DMT was in a vape pen and was dosed at .25%; it wouldn't be as dramatic as the full dose. I was relieved. After setting up, we set intentions, discussed the mechanics and protocols, and grounded ourselves in the space. After prayers and smudging, I was ready.

I sat cross-legged with a drum in front of me, hoping that at a quarter dose, I would be able to drum while I journeyed. She handed me the vape pen, and I took a long drag, deeply inhaling as much DMT as possible while holding it in for as long as I could, ignoring the burning in my lungs.

Soon, I was straddling both worlds. I was conscious enough of this world to pick up the drum. I beat the drum at the familiar shamanic rhythm of 180 bpm, the Theta-producing rhythm. I was also aware that I was channeling or speaking light language. I was conscious of being in another realm, like an in-between place where I could speak to the creator.

I apologized for the past three years for being stubborn, for the harm I had done, for doing it my way, for not seeking greater partnership, or for not implicitly trusting the divine or my guides as I should have. I committed to starting over, trusting the divine, trusting my guides, and doing everything in partnership going forward.

While I was in the medicine, I recalled that Joan had given me two homework assignments to complete after my formal apprenticeship, but I still hadn't completed them. One was a vision quest, and the other was a pilgrimage to my place of power. I had been so engrossed in the actual work and in doing things my way that years had gone by without me prioritizing these final two homework assignments. I was going to complete both during this sabbatical.

The following day, I canceled ninety-nine days of appointments and started my sabbatical. I also corrected a glaring problem—my spiritual practice; I had utterly abandoned it. I wasn't praying, meditating, or practicing any of the tools that I often recommended to others. I was a hypocrite. I listed everything I could think of that shaped a stronger connection with creation and raised my vibration. That list included:

- Prayer
- Meditation
- Mindfulness

- Breathwork
- Grounding
- Tapping
- Art
- Dance
- Gardening
- Divination
- Reading
- Cold plunges
- Exercise
- Journaling
- Crafting as a hobby

I also committed to stop using cannabis or psychedelics between now and the end of the sabbatical.

My priority was to see my birth mom. My intention for seeing her was to help me heal my wound with her and the abandonment. I suggested that we do psilocybin together as a way to discuss the events of my birth in a state of neuroplasticity and change the energy around it. The timing wasn't right, but the universe was watching out for me anyway. Walking through her home, I saw a shaman drum on her shelf.

"What is this?' I asked, shocked to see she had a shaman drum.

"Oh, that?" She asked.

"Just something I made about 40 years ago when I was in my pagan phase." She continued.

"Can I hold it?" I asked.

"Yes," she responded.

I picked it up, noting that it hadn't had bear grease in a very long time and was dry with a single crack near its frame.

"You know, we could get some leather oil and bring it back to life while I am here," I suggested.

"Oh yeah?" she asked.

"Yeah, it wouldn't be that hard; it really needs some oil or grease."

"Why don't you just take it?" She asked.

That night, I sat on the beach looking out over the Pacific Ocean, watching the sunset. Despite not being able to heal the way I had hoped or having an opportunity to show my birth mother my ways of using the sacred fungi, the universe found a way to give me a small degree of validation. It showed me that, in many ways, I had always been watched over and that the shamanic tradition that typically runs through bloodlines had, in fact, remained intact despite my birth mother being a hardened atheist.

A week after my trip to Washington, I conducted my vision quest or Utiseta in the Nordic tradition. On the morning I chose to begin, two days before the full moon, I met at a friend's house. He was on our staff, a former Army medic, and would act as the firekeeper. The firekeeper is a sacred role, the person in charge of not letting the fire go out and holding the space for the duration of the Utiseta.

We prayed, smudged, and lit the fire. Then, I was marked with ash with various bind runes for protection, strength, and courage. We drove to the mountain base where I was to conduct the Utiseta. We walked in darkness until we were at the designated spot by 5:30 a.m., a half an hour before sunrise. I set out what few things I had—a ground cloth, a blanket, a drum, a crystal, and a journal. We marked the 10-foot circle that I was to stay within, said more prayers, did more smudging, and he departed just as the sun was about to rise. He would be the only human who knew where I was and my only approved connection to the outside world. I was to sit under that tree, praying and meditating without food, water, shelter, fire, or phone for three days to receive my vision.

For three days, I was rewarded with over forty pages of notes in my journal, and I experienced great miracles with wildlife, the elements, spirits, and my guides. I had one of the most magical three days of my life. For example, at one point, my heart felt some sharp pains. I was concerned for my safety as I knew that the lack of electrolytes was a risk to my health, or potentially fatal. I prayed for comfort, strength, and courage. Just as I did, a red-tailed hawk swooped from east to west right in front of me and then cawed sharply. I felt like I was being watched over.

An hour passed, and I heard footsteps approaching my position. I was freaked out, feeling like someone was about to stumble into my sacred space and ceremony.

I sank under the tree hoping not to be seen, and heard a soft, "Hey brother, you doing okay?"

It was my friend who came to check on me.

"Yeah, yeah," I replied.

"Okay, well, I was working in the yard, and a red-tailed hawk dive-bombed me, so I knew I had to check on you." He said from outside the circle, honoring his role as the firekeeper by checking on me and not violating the circle's sanctity.

I told him about the pain in my heart, my prayer, and the red-tailed hawk I had seen.

He asked me some questions and assessed my health. We both agreed that I would be fine, and he left, not to return until 5:30 a.m. on the third day. On completion of the full three days, he reappeared, crossed the circle, and lit a new fire. As the sun rose, I burned papers containing writing of things that I no longer wanted in my life, like stubbornness or pride. One by one, I read what it was and why I wanted to be done with it. I gave it up to the universe to have them return as ash to the earth and smoke to the air. I spoke to each—why

I had been struggling with them. This lasted for almost an hour as I gave up 25 individual things.

After that, we prayed one more time and closed the ceremony. With the fire still burning, we enjoyed warm bone broth, goat milk yogurt, crackers, and boiled eggs while I shared my experience with him. One of the two remaining homework assignments was complete; one left —the pilgrimage to my source of power.

I thought a lot about the destination of my pilgrimage. Joan said I was to find "the source of my power." When she gave me that homework assignment, she let me off the hook and said I didn't need to travel to a physical place. But I wanted to experience these locations, temples, cities, and burial mounds in person. One week later, I departed for Stockholm, Sweden. I set the limitation of being home at the end of the 99 days and paying for the travel from what I could earn shamanically along the way, and the kindness of others.

For the next eight weeks, I wandered around Scandinavia, intending to walk in my past life's footsteps, visit ancestral lands, visit family, and meet a Scandinavian shamanic mentor. Most of all, I set out to prove or disprove whether or not the universe had my back, that I could trust the universe to provide, or if the universe was even aware of me.

One of my first destinations was Uppsala. Uppsala is one of Sweden's most important historical and archaeological sites, known for being the center of early Viking spiritual and political power. It has been a focal point of Scandinavian paganism and Norse mythology. I approached it on foot on a trail reported to lead directly there. As I walked the trail, I began to feel a strong sense of emotion. It felt like home; it felt natural and familiar. At one point, I dropped into a vision. I could see my past life's family walking with me. I saw a wife, a daughter, and a son. We were all walking in a caravan of people, wagons, and animals. The spirit was merry and celebratory. At a certain point, I had an overwhelming feeling that around the next corner, I would see Uppsala.

I could barely contain my excitement. As I rounded the corner, sure enough, there in the distance were the burial mounds rising on the horizon. For the next three days, I would explore burial mounds, offer sacrifices during the midnight Solstice at an altar found at the location of the tree of life—Yggdrasil—and celebrate Midsommer, the largest holiday in Sweden. I felt like I was off to a good start.

Throughout these first few weeks, I would spend time in Stockholm offering rune castings, busking on the corners, exploring ancient archaeological sites, skinny dipping under the midnight sun, and immersing myself in the Scandinavian culture and history. Anytime someone asked if I was on a holiday, I told them I was on a shamanic pilgrimage. Being direct relieved me of having to explain in greater detail later in the conversation.

Having conversations about shamanism with everyone I met along the way helped with generosity and making connections. A connection would lead to another introduction, conversation, and connection. Before long, a friend had introduced me to a couple, who introduced me to a tour guide and someone who "knew about those things," who introduced me to a woman in Denmark who taught about Seidrs and directed me to a man named James.

James was 82 years old, had been practicing shamanism for over 45 years, had been teaching it for 40 years, had lived in Denmark and Sweden, and had studied with and was friends with Michael Harner, Sandra Ingerman (a couple of real leaders in the community, whose works I had read), and Thich Nhat Hanh (another personal hero of mine). The cherry on top was that he was a Vietnam vet, a soldier. This is who I had been searching for over three years. He was teaching two courses during the 2nd and 3rd weeks of July, but sadly, both classes were filled and had a waiting list. I decided to send an email anyway. I wrote a lengthy email explaining my unique circumstances, asking for a phone call. James agreed to a phone call on July 2nd and, after our conversation, permitted me to attend both courses.

While studying with James, it was revealed that three new spirit guides were coming on board for me. I received a red-tailed hawk, a mouse, and a dragon. Dragon was the most difficult for me to connect with. I asked a woman in my class from England where the best place to connect with dragon energy was. She said I should visit Bristol, where she and her husband lived. From there, I could visit Stonehenge, Avebury, and Glastonbury. I had no idea what was in store.

On the first day in England, her husband took me to Avebury. Avebury Henge is one of the largest prehistoric stone structures, constructed around 2500 BCE, covering about 28 acres. The village of Avebury lies within the stone circle, creating a unique relationship between the ancient and modern worlds. The site has been a focus of interest for archaeologists, historians, and spiritual seekers, with many drawn to its mysterious atmosphere and connections to ancient practices.

We parked in a parking lot that offered immediate access to the stone circles. I couldn't contain my excitement as I made my way to the nearest rock that stood nearly 10 feet above ground. As I touched that ancient stone, my entire body began to vibrate. It was as if something ancient woke inside me; some part of me recognized where I was, and it felt like home.

I hugged that first rock, tears rolled down my cheeks, and I whispered, "Grandmother, I am home, I have missed you so much."

I thanked her for remembering me, being a wisdom keeper, and being a reminder of a sacred past life.

While we walked the entirety of Avebury, I gathered wildflowers to leave as an earth altar and offering to the ancestors, the earth, and the gods, as I had done throughout the pilgrimage. After several hours, it was time to go. I was overflowing with emotion, feeling a sense of sadness for leaving, overwhelmed with the knowledge that I had spent a past life there, and overwhelmed with the joy of having yet another miracle unfold for me.

The next morning, as I left the Airbnb, I met a woman named Annie from France who asked me where I was going.

"Stonehenge," I said.

"Oh, I was just there a couple of days ago. When you return this evening, stop in and tell me what you thought." She replied.

"Will do," I said as I pulled my door shut.

Stonehenge was equal parts a Neolithic sacred worship site and Disneyland. There were so many tourists that it distracted me from the sacredness of the experience. I did my best to remain positive, pulled my runes, smudge, crystal, and pendulum from my pack, and performed my own ceremony amid the sea of tourists. It contrasted sharply with Avebury but still provided a powerful connection to the site.

That night, I knocked on her door, and we discussed Stonehenge, Avebury, and Glastonbury runes, past lives, and shamanism for several hours. I told her I was going to Glastonbury the following day. I kept feeling prompted to invite her to accompany me. I ignored the prompting because I was a strange man in a foreign country, but I couldn't shake the feeling.

As we said goodnight, I finally asked, "Would you like to go to Glastonbury tomorrow?"

"I was hoping you would ask," she said enthusiastically.

The next morning, we set off for Glastonbury. Glastonbury is renowned for its deep spiritual and mystical connections. It has long been considered a place of pilgrimage, attracting people seeking spiritual enlightenment, healing, and a connection to the divine. The town's spiritual significance is rooted in Christian, pagan, Arthurian, and New Age traditions.

The first location Annie took me to was one of the most important Christian pilgrimage sites. Glastonbury Abbey is said to be the location where Joseph of Arimathea, who buried Jesus, established the first Christian church in Britain. It is also believed that Joseph brought the Holy Grail to Glastonbury and planted the Glastonbury Thorn, a tree said to bloom at Christmas and Easter, further connecting the town to Christian mysticism.

Annie walked me into the basement of the Abbey, sat down, and began to pray and meditate. She hadn't mentioned this, but I knew how to read the room, so I, too, started to pray and meditate. As I did, I could see pink and purple skin made up of scales. My dragon was beginning to reveal itself to me. I didn't mention anything to her but remained reverent as the magnitude of the moment came to my awareness.

Next, we visited the burial place of King Arthur. According to one version of the myth, Glastonbury is Avalon, the mystical island where Arthur was taken to be healed after his final battle. Additionally, Glastonbury Abbey is claimed to be the burial site of King Arthur and Queen Guinevere. The Holy Grail, the cup of spiritual enlightenment in Arthurian legend, is said to be hidden somewhere in Glastonbury, adding to its mystical allure. Again, we paused, removed our shoes, and kneeled before his grave.

Glastonbury has become a hub for New Age spiritual practices, attracting people from various belief systems. As we wandered, we found a small square at the end of an alleyway called the Divine Goddess, which was significant, I would later discover.

Next, she took me to the Chalice Well, which is a sacred spring that has been revered for thousands of years. The water, rich in iron, has a reddish tint and is believed to have healing properties. The well is often associated with the Holy Grail and is considered a symbol of the divine feminine and the life-giving force of the earth. Pilgrims visit to drink and soak in its water, meditate in its tranquil gardens, and seek healing.

Again, I took my lead from her. Silently, we entered the gardens and removed our shoes before sitting beside one of the reflecting pools. Again, Annie began to pray and meditate. Immediately, I did the same. As I did, I could see the yellow eyes of my dragon spirit guide. By now, I had realized what she was doing. She acted as a shamanic tour guide, sharing the wisdom and insight she had gained the week prior and selflessly guiding me to my dragon.

She stood quietly and led me deeper into the property, stopping at a fountain. There, we drank the water, filled our water bottles, and washed our hands and faces in the sacred waters. Without saying a word, she led me deeper still into the garden. Under a beautiful tree with benches all around, we stopped at the source of the water. Again, she stopped, and we resumed praying and meditating. This time, my dragon showed me her heart. It was made from thin glass, no thicker than the glass of a light bulb.

"How can this be?" I questioned.

"What do you mean? My dragon asked.

"Glass that thin is fragile; that's silly," I said arrogantly.

"That's the point: the divine feminine has fragile hearts, and it is time you learn how to treat them," Dragon said softly.

We sat in silence as tears rolled down my cheeks. Annie reached out and held my hand, comforting me.

"We have one more stop," Annie said.

"Where to next?" I asked.

"There," she said, pointing to the highest point as far as the eye could see.

Glastonbury Tor, a hill with the iconic ruins of St. Michael's Tower at its summit, is perhaps the most prominent spiritual landmark. It has been associated with ancient pagan rituals and is considered by many to be a powerful energy center or "earth chakra." Some believe it is a gateway to other realms or dimensions, and the ley lines (mystical alignments of land energy) are said to converge here, amplifying its spiritual energy.

After considerable hiking, we reached the top and stumbled into a trio of musicians singing earth songs. We sat down and enjoyed the free concert. After they concluded their singing, we moved outside to watch the sunset. Before long, we resumed meditating and praying. This time, my dragon, for the first time, appeared to me as a small, pink, and purple dragon with yellow eyes and a fragile heart. I watched her fly in acrobatic style across the sky before she flew toward me, stopping in mid-air in front of me.

"What is your name?" I asked.

"Majestic." She replied.

"Thank you for revealing yourself to me today."

"You're welcome. I look forward to working with you." She said.

"Me too," I replied.

We wandered back into town, found an Indian restaurant, and spoke about the day. Annie confessed she had been "told" to guide me if I asked. She had heard a sacred call and, without question, fulfilled it, giving up a day of her time to do so. I told her I had learned about my dragon a few weeks ago and came to the Bristol area to connect with it. I explained to Annie that I knew what she had been doing by the second stop. I explained that as I meditated, different parts of my dragon were being revealed to me.

As I finished telling her about Majestic, tears again were rolling down my cheeks, and that is when she said that Glastonbury sat on two converging ley lines, both known for dragon energy. That's when it all came into focus: dragons, King Arthur, Merlin, Annie acting on a prompting and being my shamanic tour guide, my dragon—this had all been orchestrated by the universe.

I ended my pilgrimage in Ireland to visit Neolithic sites and more ancestral lands on my adoptive family's side. I planned to visit the Hill of Tara, New Grange, and Loughcrew. They were described as "passage tombs."

I decided to stop at the visitor center in town and see if there was anything else worth seeing.

A woman working there offered me a tour and asked, "Are you a tourist?"

"No, I am on a shamanic pilgrimage and just came from Loughcrew," I replied.

"Really?" she asked in a "tell me more" way.

I told her about the past seven weeks, how I had visited three passage tombs in the past three days, and some of the things I had experienced.

"Come with me," she said as she spun around.

I followed her back to her desk.

"I am going to give you a number; call it this evening."

"Okay," I replied with a sense of confusion.

"Tell them I told you to call them and ask for the key to Four Knocks." She continued.

"What is Four Knocks?" I asked.

"It is another one of these sites, but it isn't open to the public." She explained.

"Oh wow," I said, feeling speechless.

That evening, I called the number and spoke to an elderly lady. I planned to meet her in the morning. I was at her home by 10 a.m. After a short conversation and a few instructions and directions, I had the key in hand and walked the last few miles to the tomb. I was still in disbelief. Was I just given the key to a privately managed Neolithic passage tomb that I could go inside and do whatever I wanted?

I started walking the road to the tomb, picking flowers and gathering items for my altar. I was eager to arrive but wanted to savor the moment as well. I almost missed the path nestled between the two rows of blackberry bushes and the small hand-painted sign. I turned from the road onto the path and followed it briefly until the raised mound was in plain view. It was considerably smaller than the others, and agricultural fields bordered it on all sides. Soon, I was standing at the steel door and turning the key, which looked like it was from the 19th century. I removed my shoes and entered the tomb quietly and gently, stopping in the center to take it all in. After a moment, I laid my pack down and began to unpack. I pulled out a candle, my runes, a pendulum, and my rattle.

In the spiritual space, there is the phenomenon of receiving a "download." When a person receives a download, their ears begin to ring, like tinnitus, but with the added presence of information being downloaded into their consciousness. Most of my downloads were subtle or gentle. This one was closer to painful. My left hand reflexively moved to cover my left ear to prevent my eardrum from exploding, but I caught myself and resisted the temptation to cover up and allowed the download instead.

My intuition was confirmed about these passage tombs. While human remains have been found in each of them, they are not just tombs. I was being informed about these structures. What I was told was that these were shamanic workshops. They all had several things in common. They were all built from stone; they were all built like a dome. They all had a single point of entry and exit. They all had a central spot in the center. They all had enclaves built into the other three cardinal directions, and they all had shamanic carvings in them. Some of the most fascinating carvings were wavy lines carved on top of each other. A tour guide in one of the tombs suggested it was most likely a lost language. I am unsure how they arrived at this conclusion, but there were only six lines matching length and waves. I am uncertain if they were suggesting these were letters, words, or sentences, but none of that felt plausible. But now, sitting alone in this tomb, looking at the carvings, acting almost like a fireplace mantel over the enclave area, it was evident that they indicated the sound frequency used by the shamans to heal their tribe.

Sound healing is a well-documented ancient healing practice gaining modern popularity. Joan had already taught me about how illnesses have their own frequency and energy waves can heal these illnesses. Sound healing operates like tuning forks, where the proper frequency is introduced and attunes the body to the proper frequency, allowing for healing. The solid rock construction and dome shape allowed for excellent acoustics, directing the sounds, chants, drumming, and singing to the patient in the center of the structure. This information also accounts for the low number of remains found in them. Most tombs had only a few remains or as many as a couple of dozen remains. If these were "passage tombs" for the burial of prominent tribe members and had been used for several hundred years, would they not have hundreds of remains?

In traditional prehistoric shamanism, when someone was identified as having "gifts," they were taken to the shaman to become an apprentice. Anciently, these apprentice relationships lasted until the master shaman died and the apprentice took over. In some instances, this may have been many decades long. If the shaman were to be buried in what was essentially his workshop, this would account for the low number of remains found in each.

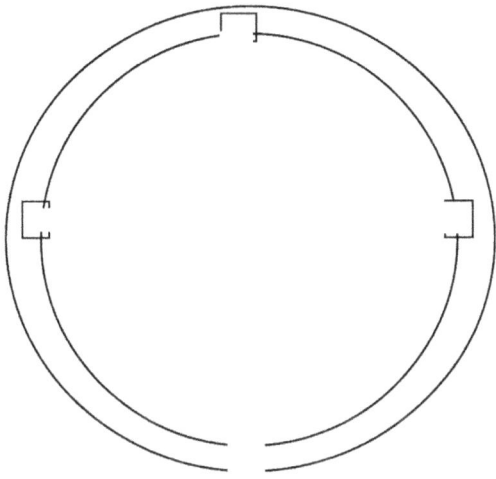

I sat inside the dimly lit tomb, feeling the energy. It felt familiar—like home; it felt natural being there. The first thing I wanted to do was to pray inside the tomb. Following my prayers, I meditated. Next, I pulled runes and asked questions with my pendulum. Next, I experimented with sounds and frequencies bouncing off the walls. I took photos to go with my notes in my journal and made drawings. In total, I was there for about six hours.

I had found the "source of my power" on two accounts. I had walked in my ancestral lands and past life paths, visited ancient shrines and temples, performed shamanic ceremonies every step of the way, and traced my ancestry on both my biological and adoptive families. More importantly, I found that the real source of my power was trusting the universe and having a deep, habitual spiritual practice. For 99 days, I didn't miss a day of journaling, praying, meditation, and divination. Most days, I also performed several other modalities: cold plunge, art, crafts, earthing, and tapping were some of the most common. The real source of my power wasn't in Scandinavia; it was in the dedicated spiritual practice. I can say that for the first time since my very first cup of ayahuasca, I finally felt more confident than I felt apprehensive. It was time to go to work!

PART THREE
ONE MAN'S JOURNEY
TO HEAL

11

TRAUMA, TBIS, PTSD, MORAL INJURY, AND ADDICTION

"Trauma is not what happens to you, but what happens inside you as a result of what happens to you." - Gabor Maté, a renowned physician and trauma expert.

When I was spiritually handed this book, *From Green Beret to Shamanism, One Man's Journey to Heal,* I would see the journey to healing portion as having at least three layers. Layer one was the autobiographical nature of describing my journey to heal myself, and what I learned as I crawled out of that abyss I found myself in.

Layer two was about my role as Seidr and my journey to heal those in my tribe. Shamanism has always been tribal. Every known example through time and in every location has a man or woman as the tribe's healer. Historically, tribes were often bound or defined by geography, fenced in by mountains or rivers, or nomadic, and the healers knew each tribe member intimately. The healers would live on the outskirts of the tribe, maintaining a degree of separation and distance, not allowing themselves to be deeply intertwined in the daily lives of the tribe, but accessible for the healing functions they provided.

Today, our tribes are not bound by geography but by those we identify with. We use the #tribe when a group gathers for like-minded activities or like-minded philosophy. The same goes for me; I see my tribe mostly made up of the broken and lost, masculine veterans, and especially the SOF veterans. I seldom work with women unless they are friends, and I never work with family. The same ethical rules apply; a doctor would never operate on his own family.

Layer three was about sharing my transformation and journey from rock bottom to Seidr to help society at large. It is intended to be a how-to guide. Let's begin with some context before we get into the "how to."

WHAT IS TRAUMA

Before we dive too deeply into the "journey to heal" portion of the book, I believe it would be helpful to understand some basic mechanics, science, and definitions. As we explore the methods for healing, patterns emerge, patterns highlighting how each of these proven methods for healing works to correct the damage done by trauma and its resulting PTSD.

In clinical terms, trauma refers to the emotional, psychological, and or physiological responses to an event or series of events that are overwhelmingly distressing or harmful. Trauma can have significant impacts on an individual's mental and physical health.

Trauma can be categorized as:

- **Acute Trauma:** This results from a single distressing event, such as a car accident, natural disaster, or a violent attack.
- **Chronic Trauma:** This involves exposure to multiple or prolonged distressing events, such as ongoing abuse or neglect.

- **Complex Trauma:** This refers to exposure to multiple, often interpersonal, traumatic events over a prolonged period, such as repeated abuse or domestic violence.

Clinical Explanation of the effects of trauma:

- **Neurobiological Impact:**
    - **Stress Response System:** Trauma activates the body's stress response system, particularly the hypothalamic-pituitary-adrenal (HPA) axis and the autonomic nervous system. This leads to the release of stress hormones like cortisol and adrenaline.
    - **Brain Changes:** Trauma can affect brain structures such as the amygdala (involved in emotional processing), the hippocampus (involved in memory), and the prefrontal cortex (involved in executive functions and regulation of emotions). These changes can result in heightened stress responses, memory problems, and difficulties with emotional regulation.

- **Psychological Effects:**
    - **Intrusive Symptoms:** Trauma can lead to intrusive thoughts, flashbacks, and nightmares related to the traumatic event. These symptoms are often seen in conditions like Post-Traumatic Stress Disorder (PTSD).
    - **Avoidance and Numbness:** Individuals may avoid reminders of the trauma and experience emotional numbness or detachment from others.
    - **Hyperarousal:** Increased arousal symptoms include heightened alertness, irritability, difficulty sleeping, and exaggerated startle responses.

- **Cognitive and Emotional Impact:**
  - **Negative Beliefs:** Trauma can lead to persistent negative beliefs about oneself, others, or the world. Individuals might feel a sense of hopelessness, guilt, or shame.
  - **Emotional Dysregulation:** Trauma can impair an individual's ability to manage and express emotions effectively, leading to mood swings, anger, or depression.

- **Behavioral Responses:**
  - **Risky Behaviors:** Some individuals may engage in risky behaviors or substance abuse as a way to cope with the distressing feelings associated with trauma.
  - **Social Withdrawal:** Trauma can result in withdrawal from social relationships or difficulties maintaining interpersonal connections.

- **Physical Health Implications:**
  - **Chronic Stress:** Ongoing trauma or unresolved stress can contribute to physical health issues, such as cardiovascular problems, immune system dysfunction, and chronic pain.

- **Clinical Descriptions of Trauma-Caused Mental Health Disorders:**
  - **Post-Traumatic Stress Disorder (PTSD):** A condition characterized by symptoms such as re-experiencing the trauma, avoidance, and increased arousal.
  - **Acute Stress Disorder (ASD):** A condition similar to PTSD, but occurs within a month of the traumatic event and lasts for less than a month.

o **Trauma-Related Disorders:** Trauma can also contribute to other mental health disorders, including depression, anxiety disorders, and substance use disorders.

The Autonomic Nervous System: The autonomic nervous system (ANS) controls involuntary bodily functions and is divided into two main branches: the sympathetic and parasympathetic nervous systems. Ideally, in a person without PTSD, both systems work together to maintain homeostasis, but have opposite effects on the body.

The Sympathetic Nervous System (SNS): Often referred to as the "fight-or-flight" system, the SNS prepares the body to respond to stress or emergencies. It activates the body's resources to handle perceived threats. When it becomes activated, a person will experience the following:

- **Increased Heart Rate:** Accelerates heart rate to pump more blood to muscles and vital organs.
- **Dilated Pupils:** Enlarges pupils to improve vision.
- **Bronchodilation:** Expands airways in the lungs to increase oxygen intake.
- **Increased Blood Pressure:** Raises blood pressure to ensure effective circulation.
- **Energy Release:** Stimulates the release of glucose and fatty acids into the bloodstream for quick energy.
- **Digestive System:** Inhibits digestion and other non-essential functions to prioritize energy for immediate needs.

In other words, your body prepares for the fight or the flight. During more than a decade of therapy, a therapist taught me how to recognize these symptoms to be able to recognize a PTSD episode coming on, or what I would later recognize as a turn in the addiction cycle. A person with PTSD will spend an inordinate amount of time in fight-or-flight.

Parasympathetic Nervous System (PNS): Often called the "rest and digest." The PNS system is responsible for calming the body and conserving energy. It promotes relaxation and recovery after stress.

When it becomes activated, a person will experience the following:

- **Decreased Heart Rate:** Slows down the heart rate to maintain a calm state.
- **Constriction of Pupils:** Reduces pupil size to protect the eyes and manage light exposure.
- **Bronchoconstriction:** Constricts airways to normal levels after stress.
- **Decreased Blood Pressure:** Lowers blood pressure to promote relaxation.
- **Digestive System:** Stimulates digestion and other restorative functions to aid in the recovery and maintenance of body systems.

Both systems work in tandem to ensure the body responds appropriately to various situations and maintains balance. The SNS prepares the body for immediate challenges, while the PNS supports long-term health and recovery.

ANCESTRAL TRAUMA

Ancestral trauma, also known as intergenerational trauma, refers to the transmission of traumatic experiences, emotional wounds, or unresolved psychological distress from one generation to the next. It can be passed both biologically and culturally. It can influence behavior, mental health, and stress responses in descendants, but awareness and healing practices can help break the cycle of trauma. Psychological transmission is when emotional and psychological trauma experienced by ancestors, such as war, slavery, genocide, or abuse, can influence how future generations perceive the world, respond to stress, or form relationships.

Epigenetics is the science discipline that studies how trauma leaves biological marks on DNA, altering gene expression. These changes can be passed down to subsequent generations, making them more susceptible to stress or mental health challenges. In 2013, researchers Brian Dias and Kerry Ressler at Emory University proved that trauma could be passed down through generations in animals. The experiment exposed male mice to a specific scent paired with a mild electric shock. Over time, the mice learned to associate the scent with pain and showed fear responses when exposed to the smell alone.

Subsequent phases of the experiment involved exposing the offspring of these conditioned mice. Despite never being exposed to the scent or the shocks, they also exhibited fear and heightened sensitivity to the same smell. Even the grandchildren of the original mice showed similar fear responses despite not having direct exposure to the trauma. The study found that this behavior was linked to changes in the DNA of sperm cells from the conditioned mice. Specifically, the genes responsible for smell receptors that detect acetophenone were altered, leading to increased sensitivity in the next generations.

This study shows that trauma can be transmitted across generations through changes, suggesting that experiences like fear and stress can have biological impacts on future generations, supporting the concept of ancestral or intergenerational trauma.

Trauma is also passed down through behaviors, beliefs, coping mechanisms, and cultural attitudes. Families or communities shaped by past trauma may unintentionally perpetuate patterns of fear, anxiety, or mistrust. Recognizing and addressing ancestral trauma through therapy, spiritual practices, or cultural healing rituals can help break the cycle of pain, allowing individuals to heal and prevent further transmission.

TRAUMATIC BRAIN INJURIES (TBIS)

To make matters worse, I had lived a hard life as a farm kid, a combat sports enthusiast, and then 27 years in the Army; I couldn't put a number on the number of concussions and head injuries I had suffered. A Traumatic Brain Injury (TBI) occurs when a sudden external force causes damage to the brain. TBIs can range from mild (like a concussion) to severe, and they can lead to a wide variety of physical, cognitive, and emotional symptoms depending on the severity and location of the injury. Treatment varies based on the severity, and long-term effects can persist, especially after multiple injuries. TBIs are typically caused by falls, motor vehicle accidents, sports injuries, or violent incidents (e.g., gunshots or explosions). In military contexts, blasts are a common cause.

Types of TBIs

TBI symptoms include headache, dizziness, confusion, fatigue, emotional changes (like depression or anxiety), chronic headaches, difficulty with speech, movement, and cognitive function, permanent cognitive deficits, and may lead to long-term disability. Multiple TBIs increase the risk of neurodegenerative diseases like chronic traumatic encephalopathy (CTE). Depending on the severity, treatment can include rest, medication, and rehabilitation (physical, occupational, speech therapy, and microdosing psilocybin).

TBIs can significantly complicate PTSD, as the two conditions often overlap, particularly in military veterans and individuals who have experienced severe trauma. The presence of both can prolong recovery and intensify symptoms. Effective treatment requires a comprehensive approach that addresses the conditions' cognitive and emotional aspects. When both conditions are present, they can interact in ways that make diagnosis, treatment, and recovery more challenging. TBI and PTSD share many symptoms: cognitive impairments, emotional regulation, increased sensitivity to stress, memory problems, difficulty concentrating, irritability, and sleep disturbances. This overlap can make determining which condition is causing specific issues challenging, complicating treatment.

Cognitive symptoms from TBI may mask or delay the diagnosis of PTSD. People with TBIs may not recognize that they are experiencing PTSD symptoms, delaying treatment and potentially worsening the condition. Simultaneously treating both TBI and PTSD can be complex. Cognitive rehabilitation for TBI may not fully address the emotional and psychological trauma of PTSD, and therapies like cognitive behavioral therapy (CBT) for PTSD might be less effective if a TBI impairs the patient's cognitive abilities.

PTSD AND MORAL INJURY

In my journey to learn how to manage my C-PTSD, it was in my nature to learn all I could on the topic. I came to understand that PTSD is often the result of the most horrific traumas that humankind can offer—war, rape, physical and emotional abuse, and abandonment, to name a few. But it is more complicated than that. We also have to account for Moral Injury. PTSD and moral injury can be viewed as siblings, but they are not the same. PTSD is a wound to the nervous system, born from overwhelming fear and threat. It shows up as flashbacks, nightmares, hypervigilance, and avoidance—the body locked in survival mode long after the danger has passed. It is fear-based and rooted in the brain's wiring, a reminder that trauma does not simply fade when the moment ends.

Moral injury, on the other hand, is a wound to the conscience and the soul. It arises when a person does something, fails to prevent something, or witnesses something that shatters their deepest sense of right and wrong. The emotions are not fear, but guilt, shame, anger, betrayal, and despair. Unlike PTSD, which is often treated by calming the nervous system, moral injury requires repair of the inner moral compass—forgiveness, atonement, meaning-making, and often spiritual or communal healing.

Together, PTSD and moral injury form a heavy burden for many veterans, first responders, and survivors of violence. One attacks the body's ability to feel safe, the other erodes the heart's ability to feel worthy or whole. Healing must therefore address both dimensions: restoring balance to the nervous system and tending to the deeper questions of conscience, trust, and spirit. Only when both are acknowledged can true repair—body, mind, and soul—begin.

Aspect	PTSD (Post-Traumatic Stress Disorder)	Moral Injury
Definition	A psychological condition caused by exposure to life-threatening or terrifying events.	A psychological, emotional, and spiritual wound caused by violating, witnessing, or being unable to prevent violations of deeply held moral beliefs.
Primary Driver	Fear- and threat-based survival response stuck in overdrive.	Conscience- and value-based conflict, involving guilt, shame, or betrayal.
Core Emotions	Fear, anxiety, helplessness, hypervigilance.	Guilt, shame, anger, betrayal, spiritual despair.
Typical Triggers	Combat, assault, disasters, accidents, abuse.	Actions or inactions in war, harming civilians, betrayal by leaders, failure to save others, breaking moral codes.
Symptoms	Flashbacks, nightmares, avoidance, hyperarousal, detachment.	Loss of trust in self/others, self-condemnation, withdrawal, hopelessness, spiritual or existential crisis.
Body vs. Soul	Primarily a nervous system injury (fear pathways, brain-body trauma).	Primarily a moral/spiritual injury (identity, conscience, worldview).
Treatment Focus	Trauma-focused therapy (EMDR, CBT, exposure), medications, somatic healing.	Moral repair: forgiveness work, atonement, spiritual counseling, rituals, peer/community support.
Can Occur Without the Other?	Yes (e.g., survivor of a car crash has PTSD but no moral injury).	Yes (e.g., soldier feels guilt over moral choices but no PTSD symptoms).

Life is full of trauma; we all experience trauma. This is not meant to sensationalize or minimize our traumas; it's just a matter of fact. My belief is that one of the grand designs of life on Earth is to come here to gain experiences that only trauma can offer us, but more on that later. Following a traumatic event, we all experience post-traumatic stress. If the trauma is severe enough or beyond our capacity, and we lack the knowledge, skills, or opportunity to process it from our lives, trauma will metastasize into a disorder. We all have the potential to arrive at some level of PTSD throughout our lives.

I recognize that PTSD is a loaded term. Many of my colleagues remove the "D" from PTSD to lessen the stigma of the condition so that it is less of a disorder. By and large, I agree with the effort to lessen it, to normalize it, to take the shame of it away. For simplicity, I am choosing to stay with the mainstream and widely recognized acronym PTSD. I am choosing to remain clinical in my approach.

There should be no shame or fear in PTSD. After all, we all have had trauma, and many are, in fact, suffering from PTSD without being diagnosed. Unless we learn how to remove the effects of the stress from post-traumatic events, we all have a degree of disorder. Unless we remove the energetic impact of the trauma we experience, that energy and its ill effects will build up in our lives and will manifest in our beliefs, actions, and bodies, and maybe most debilitating—in our minds.

Think of healing as maintaining a working oil filter on our cars. As long as we keep a properly functioning oil filter on our car, the filter will remove the impurities and harmful substances that can harm the complicated machine that is an automobile. If you never changed your oil filter, it wouldn't be long until your engine experienced catastrophic failure. It is the same with us. When we fail to remove the trauma-derived ill effects of life from our bodies and minds, we begin to experience catastrophic failure.

I also want to make a very deliberate effort to point out that PTSD doesn't come with a one-size-fits-all set of rules. How I react to certain traumatic circumstances will be different than how I respond to others. Some may affect me, while others may not even raise my awareness. I may also react differently to a particular traumatic episode without regard early in my life, while the same event may cause me severe trauma later in life. And what didn't cause trauma one day may cause me trauma tomorrow. We can't compare ourselves to others, their responses to trauma, or their circumstances. What affects someone else, again, may not even register with me, or just the thought of it may bring debilitating fear. More plainly, it is not our place to judge others.

In this journey we call life, we all will experience an almost unfathomable amount of trauma. How each of us responds to and processes our traumatic events is a profoundly personal, evolving, and individual process with our own set of outcomes.

This journey to recover my mental health became a laboratory for learning for me. In fact, birth itself is a traumatic experience, and trauma imprints on your DNA. All trauma, unless unresolved, is passed on. This spiraling downward effect exponentially complicates our lives, making it harder and harder to function, sleep, remain rational, and stop reacting. With each additional trauma, we go deeper and deeper into the PTSD cycle, thinking and behaving in ways that lower the probability that we navigate traumatic experiences in a healthy way—further adding to our trauma.

I once sat in with a PTSD/Moral Injury working group[1] led by a Lakota Medicine Man and a Research PhD.

One day, while discussing PTSD, I spoke up and said, "Yeah, but it's more than just a disorder."

"What do you mean?" The PhD asked.

"It is also an addiction and a virus," I replied.

"Explain what you mean," he said with a challenging tone.

"Well, it's an addiction because when soldiers are trained, we are taught the concept of 'violence of action,' the idea that we solve problems in combat with violence."

---

1. The official name is "SOF Trauma Recovery Group."

"Okay," he nodded in agreement.

"So, if we are out on patrol and get into it with the enemy and apply 'violence of action' to the situation, and then we get to a safe place physiologically, our bodies dump serotonin, adrenaline, and dopamine into our systems. And as you know, neurons that 'fire together, wire together,'" I continued.

"Uh-huh," he nodded again, keeping up with what I said.

"So, then, every time we are in a psychologically stressful situation, my brain has been conditioned or programmed to respond with violence to get the best drug cocktail imaginable. And as an 'addict' to my own body's chemicals, I experience the same cycle of addiction."

His face had shifted from nodding and understanding to an expression of gears turning in his head as he processed this new information.

"As a former AA participant, I am all too familiar with the cycle of shame. Where I go from trigger to craving, to ritual, to use, to shame. The only difference is that my 'ritual' is violence, and my use is physiological."

He was feverishly taking notes by this time.

"It doesn't end there," I continued.

"What else?" he asked.

"PTSD is a virus," I said.

"What do you mean by a virus?" he asked.

"Yeah, it's contagious, like the flu."

"How so?" encouraging me to continue.

"Well, in ancient civilizations, warriors were segregated from the tribe until they shook off the traumas of war and killing, preventing the

PTS from becoming a disorder. Even as recently as WWII, service members returned home on slow ship carriers, allowing them to decompress, bond, grieve, and share their stories."

"Okay, what's your point?" he asked.

"I literally had the experience of leaving Afghanistan on a Wednesday and being at my daughter's softball game on a Saturday—no decompression, no bonding, no segregation from my tribe while I dealt with the effects of the PTS to prevent it from becoming a disorder. I, and more likely the NCOs, who were doing the heavy lifting, were still processing grief, trauma, and shock." I replied.

I continued, "Add in the addiction aspect, and now I am at a softball game, there is an overwhelming amount of stimulation, I'm still stuck in hypervigilance, traffic is bad, my marriage is in a fragile state because we are trying to reintegrate after me being gone for six months—do you think I am in a good place? No, no, I am not. So now, I am a stress monster; I am acting out violently and am losing my temper, yelling at my wife and kids all the time, right?"

"Okay, I see what you're saying, but how does that make it a virus?" he responded.

"Now, when dad's truck pulls into the garage from work, the kids run for their rooms, shutting their doors, the wife avoids intimacy and connection because she isn't sure who I am anymore, everyone is walking on eggshells around me, and has become hyper-vigilant—hyper-vigilant toward how to behave around me. Isn't that the 'D' for 'disorder' in PTSD?" I asked.

"Oh." He replied

"Aren't my wife and kids now behaving in an altered way? Have I not passed off a version of, or a lesser form of PTSD, to the closest family members? Is that not a virus?"

He responded with a gentle nod in the affirmative.

Western civilizations required warriors to undergo a period of isolation or purification after returning from war before reintegrating into the tribe. This practice was often done to cleanse the warriors spiritually and emotionally, allowing them to transition from a state of battle to a peaceful communal life. A few prominent examples include Native American tribes, such as the Lakota and Cheyenne, who practiced isolation and purification, such as sweat lodges, to cleanse their bodies and spirits of the bloodshed.

The Maori warriors were required to bathe, fast, and use karakia (incantations). These practices were believed to remove the tapu (sacredness) associated with battle. In ancient Greece, returning soldiers participated in purification rites to cleanse themselves of the miasma (ritual pollution) of killing. These rites often included offerings to gods and bathing in sacred waters.

Zulu warriors in Africa were required to undergo a period of isolation and perform rituals, including drinking special medicinal mixtures, to rid themselves of the "pollution" from battle. These practices reflected a deep understanding of war's psychological and spiritual toll, offering space for warriors to process their experiences and reintegrate into their communities.

Whereas I found myself departing war zones and sitting in an Applebee's in Fayetteville, NC, 48 hours later. My experience is not unique. None of the GWOT Veterans were allowed the time and space to heal from the horrors of war before being expected to reintegrate back into society, only exacerbating the PTSD.

This is why PTSD is so insidious. It isn't just a disorder. Sure, it is a clinical disorder because it significantly disrupts a person's psychological and emotional functioning. The term "disorder" indicates that the condition meets specific diagnostic criteria and is recognized as a significant mental health issue.

The criteria for a PTSD diagnosis require that individuals exhibit a range of symptoms that persist for over a month and cause significant impairment in daily functioning across various areas of life, including work, relationships, and personal well-being. As a disorder, it can lead to difficulties in managing daily responsibilities, maintaining relationships, and achieving overall life satisfaction. Symptomatically, people with PTSD often experience intrusive memories, flashbacks, or nightmares related to the traumatic event.

People may avoid reminders of trauma and experience numbing emotions or detachment from others. Other symptoms include heightened alertness, irritability, difficulty sleeping, and exaggerated startle responses. Negative changes in mood and cognition are also included. These include persistent negative beliefs about oneself or others, feelings of guilt or shame, and difficulties in remembering aspects of the trauma.

Additionally, we must consider that with every turn of the 'addiction cycle,' we add in more trauma and pass that on to others. Then, we must factor in the number of us who are undiagnosed or untreated. We also can't ignore those who use harmful, destructive methods to cover up the pain of the trauma or as a coping mechanism, leading to further addiction, disconnection, and added trauma. Who can blame them?

When my PTSD was in full swing, I would go days with little to no sleep. I was both overwhelmingly desirous of some quality rest and fearful of the demons I would face in the form of nightmares. I would have anxiety over the very thought of sleep. The only thing that would work was Jameson whiskey or Ambien sleeping pills, sometimes combining them. Without proper techniques or coping mechanisms, a person with PTSD and an addictive nature often finds that accompanying addictions become a downward spiral, frequently leading to suicide. When someone chooses to end their life by suicide, there is an exponential amount of additional trauma passed onto those left behind, further adding to the spiral's speed and depth in tight-knit communities like the Veteran community. This is why you can see a streak of suicides run through a group. When we examine PTSD in its entirety, it is overwhelmingly evident that we are facing a problem of epidemic proportions.

Now, let's talk about antidepressants and their effect on the brain.

I want to be very clear: I am not opposed to antidepressants; sometimes, they are a much-needed retreat from our depression and anxiety. However, they are not perfect either.

SSRIs (Selective Serotonin Reuptake Inhibitors) and SNRIs (Serotonin-Norepinephrine Reuptake Inhibitors) can also have some unintended effects that make it difficult for your body to regulate and access serotonin and dopamine effectively over time. They increase serotonin or norepinephrine in the brain, causing the body to downregulate receptors (reduce the number of receptors' sensitivity). Over time, this adaptation can make it more challenging to utilize naturally occurring serotonin and dopamine effectively once medication is stopped. Almost as if the resulting addiction is by design by Big Pharma? Antidepressants reduce the motivation, reward, and pleasure that are mediated by dopamine. They also increase norepinephrine, impacting dopamine levels because norepinephrine and dopamine share some pathways.

While the medication reduces negative emotions (like anxiety and depression), it also dampens positive emotions, as dopamine's role in the reward system is affected. I always said that antidepressants felt like a chemical lobotomy. By artificially altering the balance of serotonin, it is harder for the brain to regulate its serotonin and dopamine production when the medication is withdrawn. Long-term use of SSRIs or SNRIs also leads to dopamine depletion due to chronic serotonin elevation. Since dopamine is responsible for motivation, pleasure, and focus, a decrease in its availability can lead to apathy, fatigue, and other symptoms that may mimic or exacerbate depression.

233

ADDICTION

People become addicted due to a combination of factors, including genetics, environmental influences, and mental health conditions. In AA, I learned the phrase, "Genetics load the gun, and circumstances pull the trigger." My half-brother died from a heroin overdose. It isn't a stretch of the imagination to conceive that I was born predisposed to addiction.

Addiction often develops to cope with stress, trauma, or emotional pain or to try and fill the God-sized hole that can't be filled. It's important to note that anything can become an addiction. Substances or behaviors that trigger the brain's reward system release dopamine, creating feelings of pleasure or relief. Neurons that fire together wire together. When I take in a substance or perform an act that releases dopamine, those neurons associate and wire together. Over time, this can produce an addiction to almost anything. This will naturally lead to tolerance, requiring more of the substance or behavior to achieve the same effect, and dependency, where the person struggles to function without it. These factors make it difficult to stop using the substance or behavior despite harmful consequences, leading to a cycle of addiction.

In a study referred to as "Rat Park," conducted by psychologist Bruce Alexander, rats were given access to two types of water: plain water and water laced with narcotics like morphine or cocaine. The isolated rats frequently chose the drug-laced water, leading to addiction and, often, death.

However, when the rats were placed in a stimulating environment called "Rat Park"—an ample space with other rats to socialize with, toys to play with, and room to roam—they showed much less interest in the drug-laced water and mostly consumed plain water. Even rats previously addicted to narcotics reduced their drug consumption when placed in this enriched environment. This experiment suggests that social isolation and lack of stimulation were significant contributors to addiction.

If you feel like you are the only one dealing with your demons, it causes isolation and separation, increasing the likelihood of addiction. Trauma, PTSD, TBIs, and addiction all lead to one's life spiraling out of control at an exponential rate.

12

DRUGS, MEDICINES, AND SACRAMENTS

"The difference between drugs and plant medicines is that drugs make you forget, and plant medicines make you remember everything you've forgotten." —Terence McKenna

THE SHORT-TERM PLAN: LET'S EXAMINE SOME TERMINOLOGY

Drugs

Growing up in the 1970s and 80s, I was intimately familiar with Vietnam veterans. Some from the town I grew up in, one in my family. Growing up in a conservative Utah town, I was only somewhat aware of what were called "dirty hippies." When we went to Salt Lake City, we might see one wearing a faded OD Greed Jacket with a peace symbol sewn on it. My dad would make a disparaging comment, and Mom would tell us to "look away."

As a young Reagan Republican in a small, radically conservative town, I fully adopted this mindset. I recall thinking to myself what an act of betrayal it was to have fought for one's country and then come back, burn flags, smoke "drugs," and "drop out." I couldn't get my mind around it—until now . . . As I mentioned above, we were pawns in the game, which lasted 18 years longer than it should have.

It turns out the hippies were right. They were right then, and they are right now. They weren't degenerates. They were wounded psychologically. They had been forced to fight an unjust war, to participate in unspeakable atrocities, and returned home to a divided country. Half hated them for fighting, and half hated them for not winning. They did what most of us would do and checked out. They banded together, formed communes, withdrew from society, and learned to heal themselves and their friends.

One of the most significant catalysts for quality cannabis was a Vietnam veteran and fellow Green Beret named Robert "Bob" Jordan. He was known for his involvement in smuggling high-quality cannabis seeds from Pakistan into the United States during the 1970s and 1980s. Jordan played a significant role in popularizing the cultivation of potent cannabis strains in the US. He was known for his connections with Afghan and Pakistani growers and for bringing back seeds that helped establish the foundation for many of the high-quality cannabis strains that are popular today.

In the early 1970s, there was a major federal campaign aimed at reducing illegal drug use and drug-related crime. However, the motivations behind this policy were primarily to suppress the counterculture movement and anti-war protests and to enforce racial discrimination. In 1971, President Richard Nixon declared a "War on Drugs," emphasizing the need to combat drug abuse and trafficking. The initiative increased funding for law enforcement agencies and stricter drug laws, establishing the Drug Enforcement Administration (DEA) in 1973. Nixon and his administration viewed drug use as a significant component of the counterculture movement, which included anti-war activists and hippies, the anti-war protesters who used drugs, and Black and Latino communities.

In an interview with *Harper's Magazine* in 1994, John Ehrlichman, an aide who served as White House Counsel and Assistant to the President for Domestic Affairs under President Richard Nixon, said:

The Nixon campaign in 1968, and the Nixon White House after that, had two enemies: the anti-war left and black people. You understand what I'm saying? We knew we couldn't make it illegal to be either against the war or black, but by getting the public to associate the hippies with marijuana and the blacks with heroin, and then criminalizing both heavily, we could disrupt those communities. We could arrest their leaders, raid their homes, break up their meetings, and vilify them night after night on the evening news. Did we know we were lying about the drugs? Of course we did.

These initiatives led to increased law enforcement, mass incarceration, and enduring racial disparities in the criminal justice system. As a veteran suffering from PTSD who went out on a limb and tried "drugs" in a last-ditch effort to heal myself, I was shocked to learn that I had been lied to. No, shocked is too kind; I was angry, I was pissed.

According to the racist initiatives by the US Government and several international treaties, many of the treatment options I am going to discuss below are considered "drugs." Illegal, harmful, and dangerous. Ironically, in some respects, this couldn't be further from the truth.

According to the Federal Government:

A Schedule I narcotic is a category of drugs classified by the US Controlled Substances Act of 1970. Must meet three criteria: 1) no currently accepted medical use, 2) highly addictive, and 3) fatal in certain doses. Some examples of Schedule I drugs include heroin, LSD, marijuana, ecstasy (MDMA), and peyote. This classification also means that these drugs are subject to the strictest regulation, and their use is highly restricted, even for research purposes.[1]

In other words, a substance must meet all three of the criteria. Let's take a look[2]:

	Medicinal Value	High Potential For Abuse	Fatal In Certain Doses
Cannabis	Y	Y	N
Psilocybin	Y	N	N
Ayahuasca	Y	N	N
DMT	Y	N	N
LSD	Y	N	N
MDMA	Y	N	N
Cocaine	N	Y	Y
Heroin	N	Y	Y
Fentanyl	N	Y	Y
Meth	N	Y	Y
Sugar	N	Y	Y
Tobacco	N	Y	Y
Alcohol	minor	Y	Y

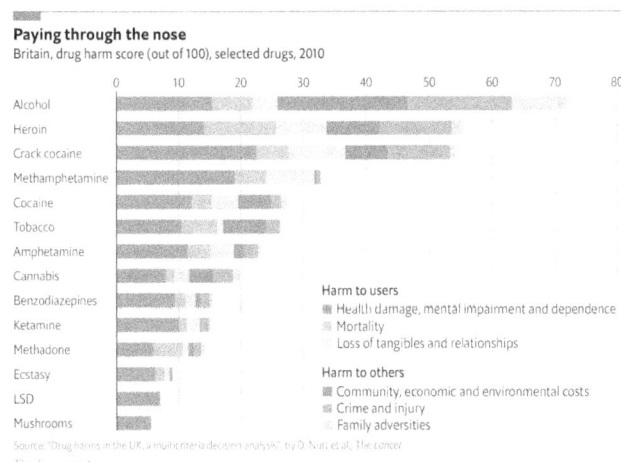

Paying through the nose
Britain, drug harm score (out of 100), selected drugs, 2010

Harm to users
■ Health damage, mental impairment and dependence
▨ Mortality
　Loss of tangibles and relationships

Harm to others
▨ Community, economic and environmental costs
▨ Crime and injury
　Family adversities

Source: "Drug harms in the UK, a multicriteria decision analysis", by D. Nutt et al., The Lancet
The Economist

The same people who keep mushrooms, naturally occurring fungi,

1. https://www.findlaw.com/legalblogs/law-and-life/drug-schedules-explained/
2. https://www.economist.com/graphic-detail/2019/06/25/what-is-the-most-dangerous-drug

out of reach for our mental health and TBI treatments are the same people who gave the green light to the opioid epidemic. Big Government, Big Business, and Big Pharma are in partnership at the cost of keeping us all sick. Let me be very clear. There are no recorded fatalities in all of human history with the following substances: cannabis, psilocybin, ayahuasca, DMT, LSD, and MDMA. Yes, there have been fatalities when there are other extenuating circumstances, such as dehydration with MDMA, but the death was due to dehydration, not the substance by itself. Even cannabis, with its higher potential for abuse, still doesn't meet all three criteria.

Our politicians are grossly uninformed as they seek to make legal decisions that harm us or put us at risk for seeking these cures.

I was once heavily involved in the advocacy space. I had to quit because it became overwhelmingly obvious that the way our political machine works is through special interest groups (the political equivalent of paid mercenaries) to pay for a politician's support. Before I quit, I had this experience with one of our senators. I explained the fallout of combat, PTSD, alcoholism, opioid abuse, and the violent arrest.

I went on to explain how an ayahuasca ceremony and psilocybin had helped me overcome those circumstances. He asked me how often I needed ayahuasca, assuming it was like a pharmaceutical and treating the symptoms—showing his own naivety. I explained that there would be one retreat, two ceremonies, and two to four doses. His mouth fell open. He quickly countered with the worn-out narrative about safety and his duty to keep harmful drugs off the streets.

I quickly countered with, "Senator, more people drowned in Utah Lake last year than have ever died from the use of psilocybin in all human history. It is impossible to OD from psilocybin."

Again, his mouth fell open as he tried to process this new information. Truthfully, criminalization creates unsafe practices. No amount of time or money has made the slightest dent in the demand for anything illegal. If there is demand, someone will be willing to supply black markets. Because of the risks associated with providing psilocybin under the laws as they stand now, those exchanges take place in the shadows. The shadows invite bad actors. If we were serious about reducing harm, we would decriminalize these activities, drag them out into broad daylight, and require licensure, regulation, and professional associations that provide training, community support, and continuing education. Think of your State Board of Realtors, but for psychedelic practitioners to truly protect the public.

Well, that's what is happening with cannabis.

To a degree, yes. But there is a difference between legalization and decriminalization. Decriminalization is modeled on the "grow, gather, give" ideology. Meaning that these naturally occurring compounds are treated like tomatoes, and that as sovereign humans, we should be free to grow them, gather them, and give them away. State-regulated cannabis dispensaries still create a legal risk for the disenfranchised. If someone can't obtain a medical marijuana card for reasons like finances, homelessness, mental health issues, or worries about being tracked, they run the risk of being arrested for partaking in lifesaving medicine that someone else with the card avoids. Interestingly, the word cartel applies here. I recently spoke to a police officer who said it this way: "One thousand out of a thousand times, I would rather respond to a call where someone is stoned versus someone who is drunk. The stoned guy will want to murder a bag of chips, and the drunk guy will want to murder everyone."

Car·tel, /kär'tel/, *noun:* an association of manufacturers or suppliers to maintain prices at a high level and restrict competition.

How are the government-run dispensaries not cartels? Do they not control the growth, distribution, and sale of cannabis without legal competition? This is the state of our elected officials, making policies and laws from a place of ignorance that keeps lifesaving plants out of our hands with the information they learned in the DARE program. It is time to abolish the DEA and its racist policies, end the war on drugs, which has yet to curb the use of drugs, and release non-violent criminals who were arrested for cannabis possession. The federal and state governments, the DEA, and the DARE program are straight-up lying to you.

PSYCHEDELICS

The word "psychedelic" originates from the Greek words "psyche," meaning "mind" or "soul," and "delos," meaning "manifest" or "reveal." The term was coined in 1956 by British psychiatrist Humphry Osmond, who was researching the effects of substances like LSD and mescaline.

Osmond was searching for a word that would accurately describe the mind-altering effects of these substances, which seemed to reveal or manifest aspects of the mind that are typically hidden or unconscious. He proposed the term "psychedelic" to emphasize the mind-revealing properties of these substances, suggesting that they could help users access deeper layers of consciousness, creativity, and self-awareness.

Osmond famously introduced the term in a letter to writer Aldous Huxley, who had also explored the effects of psychedelics. Huxley had previously suggested the term "phanerothyme" (from Greek words meaning "visible soul"), but Osmond's "psychedelic" ultimately became the accepted term. Humphry Osmond introduced the term "psychedelic" in a playful poem he wrote to Aldous Huxley. The poem reads:

"To fathom Hell or soar angelic,

Just take a pinch of psychedelic."

Osmond used this poem to emphasize the profound and mind-altering experiences that psychedelic substances can induce, whether deeply challenging (fathoming Hell) or extraordinarily uplifting (soaring angelic). The poem encapsulates the dual nature of psychedelic experiences, highlighting their potential for revealing the depths of the mind and their capacity to elevate consciousness.

The word has since become widely associated with substances like LSD, psilocybin, and ayahuasca, which are known for their ability to alter perception, consciousness, and cognition.

The contribution of psychedelics to our modern Western lifestyle cannot be overstated. In the book, *The Immortality Key: The Secret History of the Religion with No Name,* author Brian C. Muraresku details the ancient connection between psychedelics, religious experiences, and the development of Western civilization. Muraresku explains that plant-based psychedelics, particularly those derived from substances like ergot (a precursor to LSD), were central to ancient Greek and Roman societies, as well as early Christian rituals. His research is grounded in archaeology, ancient texts (including those from the Vatican), and modern scientific studies, suggesting that the use of mind-altering substances was more widespread and influential than previously acknowledged. Key points of the book:

- The Eleusinian Mysteries, a religious rite practiced in ancient Greece for nearly 2,000 years, where initiates reportedly consumed a psychedelic potion, played a foundational role in shaping early Western spirituality.
- Early Christian Eucharist rituals also involved the use of psychedelic substances.
- Psychedelics helped to shape the thinking of influential figures like Plato, who laid the groundwork for democratic and philosophical inquiry, providing insight into truth, justice, and governance.
- Breakthroughs in mathematics, logic, and scientific methodology. By challenging the boundaries of perception,

these experiences may have helped spark new ways of thinking about the natural world.

- Psychedelic use continued throughout the Renaissance and beyond, providing continuity in intellectual and spiritual progress.
- They were not merely recreational substances but pivotal tools that helped foster significant advancements in Western civilization, including the development of democracy, philosophy, mathematics, and science.

PLANT MEDICINE

Common throughout the psychedelic community is the use of the terms "medicines" or "plant medicines." This accurately describes the role these substances play in our healing. After all, they cure the illness, not just treat the symptoms. They are more medicine than many of the pharmaceuticals we are given. Many plants, animals, cacti, and fungi are all-natural remedies. One way I explain it is like this: If I were regularly constipated, a doctor might recommend that I increase the fiber-rich leafy greens in my diet. After all, that's what fiber does; that is what it is for. If we apply the same logic to naturally growing psychedelics, isn't it logical to use them for the only reason they exist, for the medicinal properties that they provide? Whether you ascribe to evolution or creation, these substances exist for a reason.

Another clue about psilocybin is its form. Dietitians recommend foods as medicines that have a visual association with the body part they are best for. For example, beets are good for the heart, walnuts are good for the brain, etc. If this holds true, examine what the mycelium network looks like. The mycelium network is the underground, thread-like structure of fungi, consisting of tiny filaments called hyphae. These hyphae weave together to form a vast interconnected web that resembles your central nervous system.

When you list the positive benefits provided by psilocybin, most of them focus on the central nervous system of the human body. However, using the word "medicine" is also problematic as it ties it to a medical procedure where only qualified medical persons should be administering or prescribing "medicines" thanks to the FDA.

SACRAMENTS

The only marginally acceptable way for someone to legally administer these substances is with the use of the Religious Freedoms Restoration Act (RFRA) in conjunction with the First Amendment. The First Amendment guarantees fundamental freedoms essential to democracy. It protects several key rights, namely Freedom of Religion, prohibits the government from establishing an official religion or unduly favoring one religion over another, and protects individuals' rights to practice their religion freely without government interference. The First Amendment is the cornerstone of American democratic principles and protects individual rights from government infringement.

The RFRA is a U.S. federal law enacted in 1993 to protect individuals' religious rights. The key purpose of the law is to *further* ensure that government actions do not burden a person's free exercise of religion. The act prevents the government from infringing on an individual's religious practices. The RFRA applies to all federal laws and actions, though state and local governments can pass their own versions of the act. It has also played a significant role in various court cases involving religious freedoms. The RFRA protects individuals' religious practices from unnecessary government interference by requiring the government to justify any actions that may restrict religious freedom with a compelling reason and to do so in the least restrictive way possible. When a person or a group of people wants to safely and legally provide these substances, the combination of the First Amendment and the RFA allows the substances in question to be referred to as sacraments.

ENTHEOGENS

The word "entheogen" comes from the Greek roots "entheos," meaning "divinely inspired" or "having a god within," and "genesthai," meaning "to generate" or "to come into being." The term was coined in 1979 by a group of scholars, including Carl A. P. Ruck and Jonathan Ott, to describe psychoactive substances used in religious or spiritual contexts. This is a key factor in psychedelics that fall within the context of entheogens having a soul, are considered entities in different forms and embodiments, and are universally accepted by almost the entire community—entities on the magnitude of that of angelic beings, no doubt about it.

The purpose of the term "entheogen" was to distinguish these substances from the more clinical or recreational connotations of terms like "psychedelic" or "hallucinogen." An entheogen is a substance that induces altered states of consciousness, primarily facilitating spiritual experiences, divine communion, or deep introspection. "Entheogen" refers to naturally occurring or growing substances used to connect with the divine, highlighting their role in spiritual practices and religious rituals. The term was created to emphasize the sacred and transformative nature of these substances, setting them apart from other uses of psychoactive drugs. For this reason, naturally occurring entheogens, more so than synthetic psychedelics, lead to a more profound spiritual awakening.

NEUROPLASTICITY

Neuroplasticity is the brain's ability to reorganize itself by forming new neural connections. Following a "hero's dose" of psychedelics, which is a high dose often used in therapeutic or transformative experiences, significant changes in neuroplasticity have been observed. Regardless of which psychedelic, such as psilocybin and LSD, all hero doses have been shown to increase neural connectivity by creating new synapses (connections between neurons). This can enhance communication between brain regions that do not typically interact.

After a hero's dose, there is often an increase in synaptogenesis, the formation of new synapses. This is partly due to the upregulation of brain-derived neurotrophic factors (BDNF), a protein that supports the growth and differentiation of new neurons and synapses. The Default Mode Network (DMN), associated with self-referential thinking and the ego, tends to become less active. This disruption allows for a more flexible and less constrained mental state, which can lead to new insights and perspectives.

Psychedelics reopen critical periods of neuroplasticity, which are times during development when the brain is particularly malleable. This reopening can enable the brain to change more rapidly, potentially leading to therapeutic benefits like reduced symptoms of PTSD or depression.

While the acute effects of a hero's dose are temporary, the neuroplastic changes can last much longer. This leads to sustained improvements in mental health, creativity, and problem-solving abilities. The enhanced neuroplasticity following a hero's dose of psychedelics is a key factor in their ability to treat mental health conditions such as depression, PTSD, and anxiety by helping individuals break out of rigid, maladaptive patterns of thinking and behavior. In summary, a hero's dose of psychedelics can significantly enhance neuroplasticity, leading to increased neural connectivity, synaptogenesis, and long-lasting changes in the brain that may have profound therapeutic implications.

"The whole is greater than the sum of its parts" was first coined by the philosopher Aristotle. My personal belief is that combining multiple approaches is better than any one approach ever will be. In other words, there is no "silver bullet," not even ayahuasca. When we look at the entire spectrum of offerings to aid and assist in treating our mental health, each approach offers distinct and differing advantages and treats varying aspects of the problem. Yes, there is a great deal of overlap in what most of them offer, but not all of them will resonate with each person, and having a broad spectrum of psychedelics and techniques to choose from offers a greater likelihood of success and greater safety.

Another of the principal advantages of psychedelics is that they can act as an "emergency brake," keeping the car from careening off the cliff. For example, it can slow or stop those in a deep depressive state or wrestling with suicidal ideation, self-medicating, failing to positively contribute to society, and suffering the fallout of these circumstances. The more their life spirals out of control, the more trauma they experience, and the more trauma they create; it's a vicious, never-ending cycle. Psychedelics offer a much-needed emergency break from the out-of-control spiral. It allows the person to get a "time out" from themselves and their addictions to see that there is a greater plan to the universe. It gives participants 30 days of neuroplasticity to form new, healthier habits and removes trauma stored in their bodies.

DETAILED OVERVIEW OF POPULAR PSYCHEDELICS

THE ENTHEOGENS

Ayahuasca (Banisteriopsis caapi and Psychotria viridis)

I love ayahuasca. She will always be my first love when it comes to psychedelics. She saved my life overnight. She showed me the reality of the universe and became my teacher and a mentor. I simply love her. And she is just that—a "she." Almost everyone who has done Aya has seen her. She appears in one of three forms, as an Anaconda, a Jaguar, or as a Latina woman, resembling most partakers will agree with these descriptions of her.

Ayahuasca is a traditional Amazonian brew made from the Banisteriopsis caapi vine and chacruna leaf for spiritual and healing purposes. The brew contains the powerful psychedelic compound DMT found in the chacruna leaf and a Monoamine Oxidase Inhibitor (MAOI) from the cappi vine, which prevents the breakdown of DMT, allowing it to be orally active. When consumed, ayahuasca induces intense visions, emotional release, and altered states of consciousness. These effects are thought to help users in several positive ways.

Ayahuasca can bring repressed emotions and traumas to the surface, allowing individuals to confront, process, and heal from past experiences. Many users report profound spiritual experiences, such as a connection to a higher power, the universe, or their inner self, leading to greater understanding and purpose. The heightened awareness and introspection during an ayahuasca journey can provide clarity on life issues, relationships, and personal challenges, promoting growth and positive change. Ayahuasca is often seen as a detox for the mind, body, and spirit, helping individuals feel renewed and rejuvenated. The experience can foster a deep sense of interconnectedness with the natural world, enhancing appreciation and respect for the environment.

Ayahuasca's effects result from the combination of DMT and MAOI, which work together to create a powerful, introspective journey. However, it should be approached with caution and ONLY used under the guidance of experienced facilitators due to its intense and challenging nature.

Indigenous tribes in the Amazon basin have traditionally used ayahuasca in their spiritual and healing practices. The exact origins are difficult to pinpoint, but archaeological evidence suggests that the use of similar psychoactive plants in the Amazon dates back at least five thousand years and possibly much longer.

These tribes have passed down the knowledge of ayahuasca from generation to generation, using it in rituals to connect with the spiritual world, heal physical and psychological ailments, and gain insight into life's challenges. The practice became more widely known beyond the Amazon in the 20th century, particularly after researchers, explorers, and spiritual seekers began documenting their experiences and the brew's effects.

In the Amazon, ayahuasca is still used in its traditional form by indigenous communities for healing, spiritual ceremonies, and rites of passage. Syncretic religious movements, such as the Santo Daime and União do Vegetal (UDV), have integrated ayahuasca into their practices. These religions, which originated in Brazil, use ayahuasca in ceremonial settings to facilitate spiritual experiences and communal worship. These groups have grown internationally, particularly in North America and Europe.

Ayahuasca has gained popularity in the West as a tool for personal development, emotional healing, and psychological therapy. Many people seek out ayahuasca retreats in countries like Peru, Colombia, and Brazil, where they participate in ceremonies led by "Ayahuasqueros" or experienced facilitators. In some places, therapists and practitioners offer ayahuasca sessions as part of holistic mental health treatment. Ayahuasca ceremonies are now conducted worldwide, including in North and South America, Europe, Australia, and Asia.

Michael Harner, an anthropologist and one of the key figures in bringing shamanism to the Contemporary Western world, had a profound experience with ayahuasca during his fieldwork in the Amazon in the 1960s. Harner drank the brew while studying the Jívaro (Shuar) people of Ecuador, and his experience deeply influenced his understanding of shamanism. During his ayahuasca journey, Harner reported intense and vivid visions, including encounters with supernatural beings and insights into the nature of reality. He described seeing dragons, celestial landscapes, and what he believed were the origins of the cosmos. These visions led him to experience a sense of connection to a greater universal consciousness and gave him insights into the shamanic worldview.

This transformative experience inspired Harner to delve deeper into shamanic practices. Eventually, he developed "core shamanism," a framework that adapts shamanic techniques for contemporary, non-indigenous contexts. Harner's work has been influential in the modern revival of shamanism, with ayahuasca playing a key role in his spiritual journey and understanding of the shamanic path.

Today, ayahuasca is used in various contexts worldwide, from traditional indigenous ceremonies to more modern therapeutic settings, extending far beyond its traditional roots in the Amazon basin. A growing interest in alternative healing practices, spirituality, and consciousness exploration has facilitated this global spread. While the legality of ayahuasca varies by country, its use continues to expand, with ceremonies often held in private or semi-public settings.

There is also increasing scientific interest in ayahuasca's potential therapeutic benefits, particularly in treating conditions like depression, PTSD, and addiction. Research studies are being conducted in various countries to explore its effects and understand how it can be safely integrated into therapeutic contexts.

Acacia

My mentor Gabrial spent seven years in Peru, apprenticing with his mentor and serving ayahuasca. It was there that he was told that he should be serving acacia. Acacia is a tree found on six continents with over 150 species, and over 75% of those contain psychedelic properties.

Mythology suggests that acacia was the actual burning bush Moses spoke to. It is acacia that I served when I was providing ceremonies for three reasons: 1) I felt that it was wrong for me culturally to provide ayahuasca, 2) I felt that acacia was safer and softer than ayahuasca, and 3) Acacia is an invasive species from the location that I get it, so it is better for the environment, not adding added pressure to the Amazon jungle. In my opinion, Acacia is Aya's daughter. There are many similarities and a few differences. Both are DMT-based, produce a roughly 6–8-hour journey, and cause purging.

DMT and 5MEO DMT

The active compound found in ayahuasca is the same active compound found in acacia and DMT. It is found in various plants, animals, and even humans. Take a moment and think about that. The active ingredient in ayahuasca, which is responsible for landing it on the Schedule 1 narcotic list, is also in your brain. It resides in the pineal gland, a small endocrine gland. The pineal gland, often called the "third eye" in mystical traditions, is located between the two hemispheres at the brain's center. It produces and regulates important hormones, most notably melatonin, which governs sleep-wake cycles.

The conclusion that DMT is produced in the pineal gland was popularized by Dr. Rick Strassman, a researcher who conducted clinical studies on DMT in the 1990s. In his book *DMT: The Spirit Molecule*, Strassman found that the pineal gland produced DMT, particularly at times of birth, death, and other significant life events, potentially facilitating mystical or near-death experiences. The specific role of DMT in the brain and its relationship to the pineal gland continues to be a subject of ongoing research.

DMT is a specific molecule with a defined chemical structure, and when we refer to DMT, we usually mean this compound. However, several related molecules within the tryptamine family are chemically similar to DMT, and these are often grouped together because of their structural similarities and psychoactive effects.

Within psychedelic space, one of the most commonly smoked associated DMT molecules is 5-MeO-DMT or 5-MeO for short. 5-MeO is a member of the tryptamine family, similar in structure to DMT but with a methoxy group attached to the 5-position of the indole ring. The most well-known natural source of 5-MeO is the venom of the Colorado River toad (Bufo alvarius), also known as the Sonoran Desert toad. The venom contains a range of psychoactive compounds, including 5-MeO. This toad's venom is often harvested, dried, and smoked to achieve its effects. 5-MeO is also found in certain plant species, although it is less common than DMT in plants. Some examples include the seeds of the Virola tree species and the plant Mimosa hostilis.

For ethical reasons, with the handling of the toads during the harvesting process and the disappearing habitat of the Colorado River toad, it is highly discouraged to smoke natural 5-MeO but instead use a synthetic form. 5-MeO can be synthesized in a laboratory setting. The synthetic form of 5-MeO is used in research and can be found in some legal psychoactive substances or research chemicals.

5-MeO is often described as having a more intense and less visually oriented experience than the standard DMT. While DMT often involves vivid visual hallucinations, 5-MeO is reported to be more focused on the dissolution of the self and profound experiential states.

When smoked, DMT produces intense and short-lived effects, typically lasting 5 to 15 minutes. DMT acts on the brain's serotonin receptors, particularly the 5-HT2A receptor, which is associated with altered states of consciousness. The effects of smoked DMT are almost instantaneous, often beginning within seconds of inhalation, leading to immediate disassociation and what most describe as an out-of-body experience or traveling to other-than-worldly realms. Users frequently describe being "launched" into an alternate reality almost immediately after exhaling.

DMT can induce profound changes in perception, sense of self, and time, often referred to as "ego death." Users typically experience vivid, colorful, and rapidly changing visual hallucinations, sometimes described as entering a different dimension or encountering geometric patterns and entities. It's common for users to report meeting beings or entities during the experience. These entities are often described as benevolent, guiding, or simply curious, though some report more neutral or even unsettling interactions. This can be liberating and disorienting, as the usual boundaries between self and the external world dissolve.

Although the experience lasts only a few minutes in real time, users often report a sense of timelessness or of being in the experience for what feels like much longer. The aftereffects of smoking DMT can include feelings of euphoria, clarity, and a sense of profound insight into oneself or the nature of reality. Some users describe it as a life-changing experience, leading to shifts in their worldview or spiritual beliefs. The overwhelming nature of the trip can be disorienting, and it's crucial to use DMT in a controlled, safe environment.

DMT is considered non-addictive and is generally well-tolerated in terms of physical safety. There's a low risk of overdose when used in its pure form, but the intensity of the experience requires careful preparation and a safe setting.

In summary, smoking DMT can induce profound and intense experiences that are often described as mystical or spiritual, with both clinical studies and anecdotal reports highlighting its powerful effects on consciousness.

Psilocybin, or Magic Mushrooms, Hero's Dose or 5 Grams

Psilocybin is a psychoactive compound found in certain species of mushrooms, such as Psilocybe cubensis. Psilocybin is found in over 200 species of mushrooms, commonly referred to as "magic mushrooms." These strains vary in potency, appearance, and growth characteristics, but they all belong to the same species (Psilocybe cubensis) and contain psilocybin as the active compound. The exact number of strains is difficult to pinpoint as cultivators continually develop new ones.

Once ingested, psilocybin is converted into psilocin, which interacts with serotonin receptors in the brain, particularly the 5-HT2A receptor. This interaction leads to altered states of consciousness, including changes in perception, mood, and cognition. Psilocybin has gained a significant amount of attention in recent years for its potential therapeutic benefits, particularly in mental health.

Psilocybin has shown promise in treating major depressive disorder (MDD) and anxiety, especially in individuals who have not responded well to traditional treatments. Clinical trials have demonstrated that psilocybin-assisted therapy can produce significant, sustained reductions in depressive and anxious symptoms. Further, psilocybin may aid in processing and integrating traumatic experiences, leading to reductions in PTSD symptoms. The psychedelic experience often helps individuals gain new perspectives and emotional release, which can be crucial in trauma therapy. Additionally, psilocybin has been shown to be an effective treatment for various forms of addiction, including alcohol, nicotine, and opioid dependence. Therapeutic sessions involving psilocybin can help individuals confront the underlying psychological factors contributing to their addiction and support long-term recovery.

Johns Hopkins University conducted a study that was as groundbreaking as it was rare for the federal government to allow the study. In this study, the Johns Hopkins University School of Medicine explored the effects of psilocybin on patients with life-threatening cancer diagnoses who were experiencing anxiety and depression related to their end-of-life circumstances. The study aimed to assess whether a single, carefully controlled dose of psilocybin could alleviate existential distress and improve the quality of life for these individuals.

The study provided compelling evidence that psilocybin, when administered in a therapeutic setting, could be an effective treatment for existential anxiety and depression in patients facing terminal illness. The results supported the potential of psilocybin-assisted therapy as a transformative approach to end-of-life care, helping patients find peace and meaning during their final stages of life. This study has contributed to a growing body of research advocating for the use of psilocybin in therapeutic contexts, particularly for those dealing with life-threatening illnesses.

Psilocybin holds significant therapeutic potential, particularly for conditions such as depression, anxiety, PTSD, addiction, and end-of-life distress. Its use should be cautiously approached, follow best practices, and ideally within a structured therapeutic framework to ensure safety and maximize benefits.

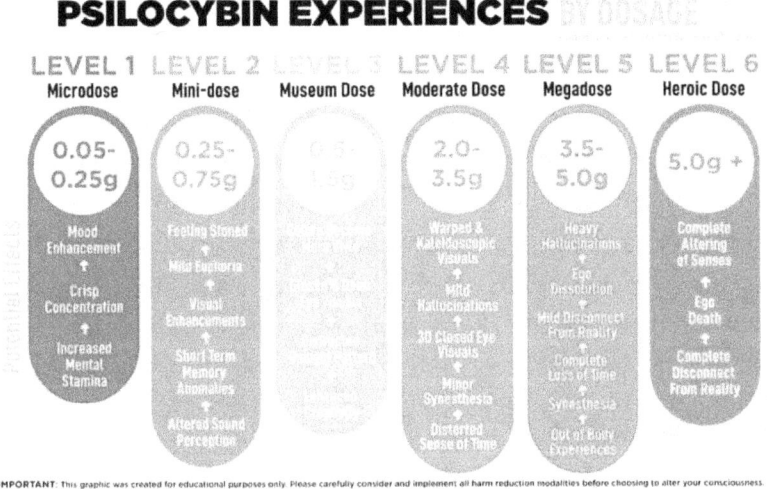

PSILOCYBIN EXPERIENCES BY DOSAGE

LEVEL 1	LEVEL 2	LEVEL 3	LEVEL 4	LEVEL 5	LEVEL 6
Microdose	Mini-dose	Museum Dose	Moderate Dose	Megadose	Heroic Dose
0.05-0.25g	0.25-0.75g	0.75-1.5g	2.0-3.5g	3.5-5.0g	5.0g +
Mood Enhancement ↑ Crisp Concentration ↑ Increased Mental Stamina	Feeling Stoned ↑ Mild Euphoria ↑ Visual Enhancements ↑ Short Term Memory Anomalies ↑ Altered Sound Perception		Warped & Kaleidoscopic Visuals ↑ Mild Hallucinations ↑ 3D Closed Eye Visuals ↑ Minor Synesthesia ↑ Distorted Sense of Time	Heavy Hallucinations ↑ Ego Dissolution ↑ Mild Disconnect From Reality ↑ Complete Loss of Time ↑ Synesthesia ↑ Out of Body Experiences	Complete Altering of Senses ↑ Ego Death ↑ Complete Disconnect From Reality

IMPORTANT: This graphic was created for educational purposes only. Please carefully consider and implement all harm reduction modalities before choosing to alter your consciousness.

Psilocybin Microdose

Psilocybin microdosing takes very small, sub-perceptual doses of psilocybin—typically about .25 grams on a regular schedule. An ideal microdose is something that doesn't alter or change perception or impairment. While not producing the intense psychedelic effects associated with higher doses, microdosing has been reported to offer a range of psychological and cognitive benefits.

Microdosing has been linked to improved mood and emotional well-being. Users often report increased positivity, reduced feelings of anxiety and depression, and a greater sense of emotional balance. Many individuals find that microdosing enhances cognitive function, improving focus, concentration, and creative thinking. It may help with problem-solving, idea generation, and out-of-the-box thinking. Microdosing can boost energy levels and motivation, making it easier to tackle daily tasks and pursue goals with more enthusiasm and drive.

Regular microdosing has been reported to reduce symptoms of anxiety and stress, leading to a calmer and more centered mental state. This can be particularly beneficial for individuals dealing with chronic stress or anxiety disorders. Some users experience greater empathy and emotional connection with others, which can improve relationships and social interactions. Although more research is needed, preliminary evidence suggests that microdosing may support mental health by providing relief from conditions such as depression, PTSD, and obsessive-compulsive disorder (OCD).

Why It Works

Even at low doses, this interaction can enhance serotonin activity, which is associated with mood regulation, anxiety reduction, and cognitive flexibility. Psilocybin has been shown to promote neuroplasticity—the brain's ability to reorganize itself by forming new neural connections. Even at low doses, psilocybin may reduce activity in the DMN, which involves self-referential thinking and mind-wandering. By modulating the DMN, microdosing may help reduce overthinking, rumination, and negative self-talk, contributing to a more positive and focused mental state. Emerging research suggests that psilocybin may have anti-inflammatory properties, which could contribute to its mood-enhancing and cognitive benefits. Chronic inflammation is increasingly recognized as a contributor to mental health disorders, and reducing inflammation could be one of the mechanisms by which psilocybin microdosing improves mental well-being.

Psilocybin microdosing offers a range of potential benefits, including enhanced mood, improved cognitive function, increased energy, and reduced anxiety. While more research is needed to fully understand its effects, psilocybin microdosing holds promise as a tool for enhancing mental health and well-being in a subtle yet impactful way.

Peyote, and Huachuma (San Pedro Cactus)

Peyote is a small, spineless cactus native to the southwestern United States and northern Mexico, known scientifically as Lophophora williamsii. It has been used for centuries by indigenous peoples in religious and healing rituals due to its psychoactive properties.

The indigenous name for the San Pedro cactus (*Echinopsis pachanoi*), particularly among the Quechua-speaking peoples of the Andes in South America, is "Huachuma" (also spelled "Wachuma"). This cactus has been used in traditional Andean shamanic practices for thousands of years. Huachuma is revered as a sacred plant medicine used in ceremonies for healing, spiritual growth, and connecting with the divine. The primary psychoactive compound in both peyote and Huachuma is mescaline, a potent hallucinogen that alters perception, mood, and thought.

Peyote is considered a sacrament central to various religious practices, particularly within the First Nations Church. In many countries, peyote is classified as a controlled substance, making its use illegal outside of specific religious contexts. In the United States, members of the First Nations Church are legally permitted to use peyote for religious purposes.

Huachuma cactus is easily purchased online or locally and can be legally possessed and grown. It is illegal, however, to consume it. Again, the ignorance and confusing nature of the law is baffling.

Peyote is endangered due to the disappearance of land where it grows, and you should not take it unless you are a First Nations person. However, there is a synthetic alternative for peyote as well.

Iboga (Tabernanthe iboga)

Iboga, derived from the root bark of the Tabernanthe iboga plant, is a powerful psychoactive substance traditionally used in Central African spiritual practices, particularly within the Bwiti religion in Gabon. Its active alkaloid, ibogaine, distinguishes it from other psychedelics due to its unique effects and therapeutic potential. Iboga affects the kappa-opioid and NMDA receptors, as well as the serotonin and dopamine systems. This diverse receptor activity contributes to its potential in treating opioid addiction and other substance dependencies. It is particularly known to reduce withdrawal symptoms and cravings, offering a "reset period" to the brain's reward systems. **This is a unique characteristic not shared by other psychedelics.**

The experience is often described as more physically and mentally intense, with effects lasting up to 18-24 hours. The duration and intensity of the experience (e.g., 4-6 hours for psilocybin, 8-12 hours for LSD) are why iboga should not be taken without proper medical attention and in a licensed medical setting. Iboga carries significant risks, including potentially life-threatening cardiac effects (e.g., QT prolongation), and is not suitable for individuals with certain health conditions. This necessitates a medically supervised environment, especially during addiction treatment.

Cannabis

I have a deep love and respect for cannabis. You and I were lied to about cannabis and its usage. While not traditionally psychedelic, it can provide psychoactive experiences under some circumstances. It can be extremely effective in relieving symptoms of PTSD, especially when working with a trained facilitator and combining somatic release.

Physical Health Benefits

Cannabis, particularly strains high in cannabinoids like THC and CBD, has shown effectiveness in alleviating various types of chronic pain (neuropathic, arthritis, fibromyalgia). CBD, one of the primary cannabinoids, has significant anti-inflammatory properties. It is thought to be helpful for conditions like inflammatory bowel disease (IBD), rheumatoid arthritis, and other inflammatory conditions. Cannabidiol (CBD) has been recognized for its anticonvulsant properties and is particularly effective for rare forms of epilepsy such as Dravet syndrome and Lennox-Gastaut syndrome. Cannabis can be helpful for individuals with multiple sclerosis (MS) or spinal cord injuries by reducing muscle spasms and improving mobility. THC, the psychoactive compound in cannabis, has been shown to reduce nausea and vomiting, particularly in patients undergoing chemotherapy. Cannabis can also stimulate appetite, which is beneficial for individuals with conditions causing anorexia or cachexia, such as cancer, HIV/AIDS, and certain eating disorders.

Mental Health Benefits

Users find that low doses of CBD and certain strains of cannabis can have calming effects, helping to alleviate anxiety and stress. Cannabis can influence mood by interacting with the endocannabinoid system, which plays a role in mood regulation. Cannabis strains, with a balance of THC and CBD, have been used to alleviate symptoms of PTSD, including nightmares, flashbacks, and heightened anxiety. Many cannabis strains, especially those high in THC or CBN (cannabinol), have sedative properties that can assist with insomnia and improve sleep quality.

Research indicates that cannabinoids have neuroprotective and anti-inflammatory effects, which may be beneficial for neurological conditions like Alzheimer's disease, Parkinson's disease, and multiple sclerosis. CBD has been studied for its potential in reducing cognitive decline and supporting brain health.

Cannabis has been used as a harm reduction tool, with some evidence suggesting it may help reduce dependence on more harmful substances like alcohol, opioids, and stimulants.[1] Cannabinoids are potent antioxidants that can help reduce oxidative stress and inflammation, contributing to overall cell health and potentially reducing the risk of certain chronic diseases.

While not universal, some users report enhanced focus, creativity, and a different perspective on problem-solving when using cannabis, particularly strains low in THC and high in CBD. Cannabis isn't without its problems, though, and is maybe the most problematic and, ironically, the most widely accepted entheogen to be legalized. Cannabis use in adolescents and young adults has been associated with an increased risk of psychosis, including schizophrenia. Research suggests that the age of first use, frequency, and potency of cannabis influence this risk.[2]

The developing brain is more vulnerable to the effects of THC, and genetic predisposition combined with environmental factors can heighten the risk. This underscores the importance of avoiding cannabis use in young people and understanding its potential risks.

SYNTHETIC PSYCHEDELICS

LSD

LSD (Lysergic acid diethylamide) has been the subject of clinical research and anecdotal reports regarding its potential positive uses, particularly in mental health, personal growth, and creativity for decades.

Recent studies have explored the use of LSD in treating anxiety and depression, especially in cases where traditional treatments have

---

1. https://www.ncbi.nlm.nih.gov/pmc/articles/PMC4604178/
2. https://pubmed.ncbi.nlm.nih.gov/30902669/

failed. In controlled settings, low doses of LSD have been shown to produce significant reductions in anxiety and depressive symptoms. The effects are believed to stem from LSD's ability to alter neural connectivity and enhance emotional processing. Many individuals report that LSD has helped them gain insights into their emotional states, leading to lasting reductions in anxiety and depression. Users often describe a renewed sense of purpose, connection, and emotional clarity after their experiences.

Studies from the 1960s and more recent research have suggested that LSD can help reduce existential anxiety in terminally ill patients. Patients often report a greater sense of peace, acceptance of death, and a reduction in fear after guided LSD sessions. There are numerous reports from individuals facing terminal illness who claim that LSD helped them confront their mortality with greater equanimity, leading to a more peaceful end-of-life experience.

While LSD has not been as extensively studied for PTSD as other synthetics like MDMA, there is growing interest in its potential to help patients process traumatic memories in a therapeutic context. The drug's ability to induce a state of heightened emotional openness may facilitate the reprocessing of traumatic experiences. Some users with PTSD report significant relief from symptoms after using LSD, describing experiences of catharsis and emotional release that have allowed them to better cope with their trauma.

Early research in the 1950s and 60s suggested that LSD could be effective in treating alcoholism, with some studies showing a reduction in alcohol consumption after treatment. More recent studies have rekindled interest in this area, with evidence suggesting that LSD can lead to a reevaluation of life choices and behaviors associated with addiction. Bill Wilson, the co-founder of Alcoholics Anonymous (AA), experienced LSD, and it played a significant role in his later life and his understanding of spiritual experiences in recovery.

In the 1950s, Wilson began exploring the potential of LSD as a tool for spiritual growth and aiding in the recovery of alcoholics. At the time, LSD was not yet illegal and was being studied by researchers for its potential therapeutic benefits, particularly in psychiatry. Wilson was introduced to LSD by Gerald Heard, a British writer and philosopher. He later participated in guided sessions under the supervision of Dr. Sidney Cohen, a prominent researcher in psychedelic therapy.

Wilson believed that LSD could induce a profound spiritual experience similar to the one he had during his recovery from alcoholism. He saw the potential for LSD to help alcoholics who struggled with the concept of a higher power or who had difficulty achieving the spiritual awakening that AA emphasized. He thought that by experiencing the ego-dissolving effects of LSD, individuals could have a direct encounter with a higher power or the divine, which might help them overcome their addiction.

Wilson's advocacy for the use of LSD in treating alcoholism was controversial within the AA community. Many members were concerned that promoting the use of a mind-altering substance, even in a controlled setting, could be seen as contradictory to the principles of sobriety that AA stood for. Eventually, Wilson stepped back from publicly advocating for LSD, but he continued to explore its potential benefits privately.

Despite the controversy, Wilson's interest in LSD reflects his broader commitment to exploring all possible avenues for helping those struggling with alcoholism. He saw LSD as a potential tool, not a replacement for the AA program, but as a means to facilitate the spiritual experience that he believed was crucial for long-term sobriety.

Although formal research is limited, there is evidence that LSD can enhance creative thinking and problem-solving abilities. The drug appears to increase divergent thinking and reduce the rigidity of thought processes, allowing for more novel and creative solutions to problems.

In fact, the discovery of the double helix structure of DNA, one of the most significant scientific breakthroughs of the 20th century, has a curious connection to LSD through one of the scientists involved, Francis Crick. Crick, along with James Watson, is credited with elucidating the structure of DNA in 1953. While there is no definitive evidence that LSD was directly used during the process of discovering the double helix, Francis Crick was known for his openness to unconventional ideas and his curiosity about consciousness and the mind.

According to some accounts, Crick used LSD in the 1950s when it was still a legal substance and was being studied for its potential to enhance creativity and problem-solving abilities. One of the most well-known anecdotes comes from a biographer of Crick, who claimed that Crick admitted to using "small doses of LSD" to aid his thought processes while working on the problem of DNA's structure. The story suggests that LSD helped Crick visualize the double helix structure, which he and Watson later famously described as resembling a "twisted ladder."

The notion that LSD could influence creative thinking and problem-solving is not far-fetched, as the drug is known to alter perception and cognition, potentially allowing individuals to see patterns and connections that might not be immediately apparent. Many artists, writers, and scientists have reported that LSD helped them overcome mental blocks and think in novel ways.

LSD has been studied for its potential to induce mystical or spiritual experiences, which are often associated with long-lasting positive changes in attitudes, behavior, and well-being. These experiences, characterized by a sense of unity, transcendence, and profound insight, have been linked to lasting increases in life satisfaction and purpose. Countless users report life-changing spiritual experiences while under the influence of LSD, describing feelings of interconnectedness, oneness with the universe, and deep personal insight. These experiences are often cited as pivotal moments in personal growth and self-understanding.

There is some evidence that LSD, even in sub-hallucinogenic doses, can be effective in treating cluster headaches, a condition notoriously difficult to manage with conventional treatments. LSD appears to reduce the frequency and intensity of headache attacks. Individuals suffering from cluster headaches often report significant relief after using LSD or related compounds like psilocybin, sometimes achieving remission where other treatments have failed.

MDMA

MDMA (3,4-methylenedioxymethamphetamine), commonly known as Ecstasy or Molly. MDMA is the pharmaceutical-grade version. Ecstasy or "X" and Molly are forms of MDMA that are cut with additional substances. MDMA has been studied for its therapeutic potential, particularly in the treatment of mental health conditions like PTSD, anxiety, and depression. Both clinical research and anecdotal evidence have highlighted its potential benefits in various contexts.

MDMA has shown significant promise in the treatment of PTSD, particularly in clinical trials conducted by organizations like the Multidisciplinary Association for Psychedelic Studies (MAPS). In these trials, MDMA-assisted psychotherapy has been found to dramatically reduce symptoms of PTSD, with some patients experiencing long-term remission. MDMA appears to facilitate the processing of traumatic memories by reducing fear and increasing trust, which can enhance the therapeutic process.

Many individuals who have undergone MDMA-assisted therapy report profound improvements in their ability to cope with trauma. They often describe the experience as allowing them to revisit traumatic memories without being overwhelmed by fear or anxiety, leading to greater emotional healing and resolution.

Research has suggested that MDMA may be effective in treating anxiety and depression, especially in individuals who have not responded well to conventional treatments. The drug's ability to enhance emotional openness, empathy, and self-compassion can lead to significant improvements in mood and a reduction in symptoms of anxiety and depression. Users frequently report that MDMA helps them achieve a heightened sense of emotional clarity and connection, which can be therapeutic for those struggling with depression or anxiety. The drug facilitates a "breakthrough" experience, where individuals gain new insights into their emotional states and relationships.

MDMA, also known as the ego killer, has been explored as a tool for enhancing emotional and interpersonal connections in therapeutic settings. It has been used in couples therapy to help partners communicate more openly and resolve deep-seated emotional issues. The drug's effects on empathy and emotional connectedness are believed to help individuals express feelings that are difficult to access under normal conditions. Many people who have used MDMA in a controlled, therapeutic context report significant improvements in their relationships. They describe being able to discuss sensitive topics without fear or defensiveness, leading to greater intimacy and understanding.

Preliminary studies suggest that MDMA may help reduce social anxiety, particularly in individuals with conditions like autism spectrum disorder (ASD). The drug's effects on empathy and social connectedness can help individuals feel more comfortable in social situations and reduce feelings of isolation. Individuals who have used MDMA recreationally often describe feeling more open, sociable, and confident in social settings. Reducing social anxiety can lead to positive social interactions and a greater sense of belonging.

MDMA has been studied for its potential to facilitate personal growth and self-understanding in therapeutic contexts. The drug's ability to reduce fear and increase emotional openness can help individuals confront difficult emotions, leading to greater self-awareness and personal growth. Many users report that MDMA has helped them gain deep insights into their behaviors, motivations, and emotional patterns. These experiences are often described as transformative, leading to lasting positive changes in how individuals perceive themselves and their place in the world.

MDMA has been explored as a tool for helping terminally ill patients cope with end-of-life anxiety. The drug's ability to reduce fear and enhance emotional connectedness may help individuals come to terms with their mortality and find peace in their final days. Some patients and caregivers report that MDMA has helped them experience a greater sense of peace and acceptance in the face of terminal illness. These experiences are often described as spiritually meaningful and comforting.

Ketamine

Ketamine, developed initially as an equine anesthetic, has gained significant attention in recent years for its potential therapeutic uses, particularly in the treatment of mental health conditions like depression, anxiety, and PTSD. Sadly, the death of Matthew Perry has amplified the call to end the use of ketamine. In my opinion, ketamine is excellent as an emergency treatment for severe treatment-resistant depression and suicidal ideation. Since ketamine isn't FDA-regulated or banned, there is a proliferation of clinics. As of 2024, there are approximately 750 in the United States. These clinics have emerged rapidly in recent years, driven by the growing use of ketamine as an off-label treatment for mental health conditions such as severe depression, anxiety, PTSD, and obsessive-compulsive disorder.

Ketamine has shown remarkable efficacy in treating individuals with treatment-resistant depression (TRD), which is depression that does not respond to standard antidepressant therapies. Intravenous ketamine infusions can produce rapid antidepressant effects, often within hours, compared to the weeks required for traditional antidepressants to take effect. Studies have demonstrated that ketamine can significantly reduce depressive symptoms, even in those who have not responded to other treatments. Many patients with TRD who have received ketamine treatments report dramatic improvements in mood and outlook, sometimes describing the experience as lifesaving. The rapid onset of relief is particularly valued by those who have endured prolonged depressive episodes with little success from other therapies.

Ketamine is one of the few treatments that have been shown to reduce suicidal ideation rapidly. In clinical trials, patients experiencing suicidal thoughts have reported significant reductions in these thoughts within hours of receiving ketamine. This has made ketamine an essential option in emergencies where quick intervention is needed to prevent suicide. Individuals struggling with intense suicidal thoughts have often reported that ketamine provided immediate relief, allowing them to regain a sense of control and stability. The ability of ketamine to "reset" suicidal ideation has been a crucial benefit for many patients.

Research has shown that ketamine can be effective in reducing symptoms of PTSD. Ketamine's ability to disrupt the patterns of negative thinking and intrusive memories associated with PTSD is thought to account for its therapeutic effects. Clinical studies suggest that ketamine can reduce the severity of PTSD symptoms, including flashbacks, hyperarousal, and avoidance behaviors. Many veterans and trauma survivors have reported significant relief from PTSD symptoms after ketamine treatment. They often describe a sense of emotional relief and a reduction in the intrusive memories that characterize PTSD, allowing them to better engage in therapy and daily life.

Ketamine has also shown promise in treating various anxiety disorders, including generalized anxiety disorder (GAD) and social anxiety disorder. Studies suggest that ketamine can reduce anxiety symptoms, potentially by modulating glutamate levels in the brain, which are thought to play a role in anxiety. Patients with chronic anxiety often report that ketamine helps to reduce their baseline anxiety levels, making it easier to cope with daily stressors. Many describe feeling a sense of calm and clarity following ketamine treatment, which contrasts with the constant worry and agitation they typically experience.

Ketamine has been used as an off-label treatment for chronic pain conditions, including complex regional pain syndrome (CRPS) and neuropathic pain. Ketamine's analgesic properties are thought to be due to its action on the NMDA receptors in the brain, which are involved in pain transmission and perception. Clinical trials have demonstrated that ketamine can reduce pain levels in individuals with certain chronic pain conditions. Patients with chronic pain who have received ketamine infusions report significant pain relief, sometimes for the first time in years. They describe ketamine as helping to "reset" their pain thresholds, providing relief that lasts beyond the duration of the treatment.

Ketamine has also been studied as a treatment for bipolar depression, particularly in patients who do not respond to traditional mood stabilizers and antidepressants. The rapid antidepressant effects of ketamine have been observed in bipolar patients, leading to significant improvements in mood and functioning. Individuals with bipolar disorder often report that ketamine helps stabilize their mood, particularly during depressive episodes. Some describe it as providing a "break" from the cycle of depression and helping them regain a more balanced emotional state.

Ketamine is increasingly being used as an adjunct to psychotherapy, particularly in the emerging field of ketamine-assisted psychotherapy (KAP). In this context, ketamine is administered in a controlled setting to facilitate deeper emotional processing and insight during therapy sessions. The dissociative effects of ketamine can help patients access suppressed memories and emotions, making psychotherapy more effective. Patients who have undergone ketamine-assisted psychotherapy often report that the experience allowed them to explore difficult emotions and past traumas in a safe and transformative way. They describe the dissociative effects of ketamine as helping to "step back" from their usual patterns of thought, enabling breakthroughs in therapy.

However, with that being said, Ketamine has its downfalls. In my experience, one of the greatest concerns is that Ketamine clinics, operating with little to no oversight, act predatory in nature. I knew of one individual who was convinced by the clinic to do a Ketamine session weekly for just over a year at the cost of $1,000 a session. It is as if they acknowledge that it doesn't hold the curative properties that ayahuasca or psilocybin do, and fall into a very expensive, pharmaceutical model.

Another chief concern is that it provides the 30 days of neuroplasticity more commonly found with other psychedelics and always found with the Entheogen category of psychedelics. The 30 days of neuroplasticity, in my experience, is one of the greatest outcomes and opportunities associated with psychedelics, allowing for the highest probability of change for those who choose this modality of healing, which leads me to my last caution with Ketamine Clinics. In my experience, most offer little to no integration. It is almost an acknowledgment that there is zero neuroplasticity or opportunity for that type of growth and change.

13
LONG-TERM SOLUTIONS

"We are what we repeatedly do. Excellence, then, is not an act, but a habit." - Aristotle

Let's talk about where the real "magic" takes place, shall we? One of the greatest advantages of psychedelics is that they allow the rider of the proverbial merry-go-round to disembark long enough to take a look at themselves in the mirror and take advantage of thirty days of neuroplasticity. It is a giant reset with an accompanying "pause button" allowing us space to address poor, destructive behaviors without the preverbal monkey on our backs. This pause gives us space to build new constructive behaviors into our lives for long-term change. One of the biggest red flags in this space is the lack of integration. If you are considering "sitting" with these substances, one of the most important criteria you can search for in your facilitator is their commitment to integration and what that integration plan looks like.

In our organization, we practice a three-phase integration process: pre, during, and post. The pre-ceremony integration includes activities such as an intake call, intake forms, diet, meditation, intention setting, journaling, and Q&A, among others. These journal entries become the blueprint for the long-term plan, which participants will continue to modify throughout the process. The blueprint also becomes the foundation for life post-ceremony.

During the ceremony, we strongly encourage our participants to bring those journals and record their insights throughout the weekend and the journey, as I did at my first ceremony. Additionally, we recognize the multiple opportunities for conversations about the ceremony, questions and answers, collective bonding with fellow participants, and other activities such as Qigong, yoga, breathwork, and meditation as forms of integration.

Post-ceremony integration is a 1-1-1 program. We offer a minimum of three integration sessions with each person. They are on day 1, week 1, and month 1 to ensure the plan is being implemented and followed during the highly valuable period of neuroplasticity; this is where the magic happens. This, in my estimation, is the greatest opportunity for change for anyone seeking healing. Ironically, it is also the greatest missed opportunity for both participants who fail to take it seriously or choose not to participate in it, and facilitators who don't offer it. Sadly, I have had my fair share of experiences that didn't provide any form of integration.

During the 30 days of neuroplasticity, we have the ability and freedom to set aside our old, destructive habits. We can form previously unattainable habits, such as a meditation practice that fosters long-term care and mental health maintenance. These substances offer us a look at both the problem and the solution, providing the time and space needed to implement a long-term solution.

Experts such as Gabor Maté, Bessel van der Kolk, and Bruce Lipton all emphasize that stress, especially chronic stress, plays a central role in the development of physical and mental illness. Our modern environments, shaped by societal pressures, unhealthy relationships, and the suppression of emotion, create a constant state of stress that undermines our well-being and leads to disease. They argue that chronic stress, especially when experienced over long periods, is a key driver of illness. Stress weakens the immune system, disrupts hormonal balance, and leaves the body vulnerable to a variety of diseases, including cancer, autoimmune disorders, and heart disease.

Stress also activates the body's fight-or-flight response, releasing stress hormones like cortisol. While this response is useful in short-term survival situations, in modern life, people often live under prolonged stress, which damages the body over time. Emotional stress coupled with unresolved trauma leads to chronic psychological stress, which eventually takes a toll on physical health. For example, people who grow up in environments where they feel they need to suppress their emotions to please others or avoid conflict often carry this emotional tension into adulthood, leading to chronic stress. They conclude that chronic stress doesn't just make people feel bad emotionally; it also plays a direct role in the development of serious illnesses. Their research shows that stress-related factors—such as emotional repression, unprocessed grief, and the inability to set boundaries— are strongly correlated with conditions like cancer, multiple sclerosis, and heart disease.

Furthermore, modern life is inherently stressful, with pressures to constantly achieve, conform, and continually suppress emotions. Our fast-paced, competitive society is a significant cause of stress-related illnesses. The lack of genuine human connection, the need to fit into rigid societal molds, and the devaluation of emotional expression create environments where stress flourishes.

According to research, chronic stress can alter gene expression, weakening the body's ability to repair itself and increasing susceptibility to illness. Changing our perceptions, beliefs, habits, and environment can reduce stress and help prevent disease. How we perceive stress is just as important as the stressors themselves, meaning that shifting our mindset can lead to better health outcomes.

All these thinkers agree that chronic stress is one of the most significant contributors to illness. Whether from childhood trauma, emotional repression, societal pressures, or unhealthy environments, stress is seen as the underlying factor that disrupts the body's natural ability to maintain health. Each emphasizes the importance of understanding the mind-body connection, particularly how psychological stress can lead to physical illness. The body responds to emotional and mental stress by releasing stress hormones and triggering inflammation, both of which play a role in the development of chronic diseases. These thinkers often critique modern environments—characterized by overwork, lack of meaningful human connection, and emotional suppression—as inherently unhealthy. They argue that societal norms that emphasize constant productivity, competition, and emotional control are major sources of chronic stress, which, over time, contribute to widespread health problems. Healing requires addressing the sources of stress and the emotional wounds behind it, reconnecting with our authentic selves, and creating environments that reduce stress and promote well-being.

The following is a list of non-psychedelic modalities. These modalities are the core of what we assist work participants with during their neuroplastic state, thirty-day period—optimistically creating new lifelong habits. They have also become the ones I use regularly—some daily, others as needed—and in various combinations to maintain my mental wellness. Interestingly, but not shockingly, they share many of the same beneficial outcomes.

All the following practices activate the parasympathetic nervous system, helping to reduce cortisol levels and promote a sense of calm and relaxation. Even in a sympathetic response, we can reverse engineer our nervous system.

Each of the following practices, in varying ways, supports holistic well-being by promoting a balanced mind, body, and emotional state. Not all of these modalities will resonate with everyone, but some of them will. I challenge you to try some, if not all, of them and find the ones that resonate with you. Play with them, try them out, what do you have to lose? Build a program that works for you, and I promise you will improve your physical and mental wellness and live a much happier life.

MEDITATION

In my estimation, meditation is perhaps the single best, non-psychedelic practice to ensure your mental health. A meta-analysis of randomized controlled trials found that mindfulness meditation programs can lead to moderate reductions in symptoms of anxiety and depression, with effects comparable to those of antidepressant medications.

I've already related my first experience with meditation, where I left my body and remembered who and what my role was. That was one hell of an introduction. Thankfully, almost none of us will have that type of experience on our first attempts. When I went on pilgrimage, I knew I would be crossing multiple borders. I also wanted to embrace a purer way to connect with the divine, so I made the decision not to use plant medicines on that trip. This left me with the dilemma of how to maintain neuroplasticity, pain management, and sleep. It was during this trip that meditation became the backbone of my spiritual practice, my mental and physical wellness. Through a twice-a-day practice of morning and evening, I was able to experience the same benefits of microdosing psilocybin and smoking cannabis.

When recommending meditation to men, I am almost universally met with resistance. I've learned that most of the resistance comes from the false perception that, as practitioners, we are trying to empty or clear our minds of thought(s). There are several reasons why we cannot halt our minds from "thinking."

One is evolutionary design. The human brain evolved to constantly scan for threats, opportunities, and patterns in the environment. Even at rest, this "DMN" remains active to simulate possible futures, replay past events, and strategize. DMN. Neuroscience shows that when we're not focused on a task, our brain switches to the DMN, which generates spontaneous, often self-referential thoughts—daydreams, memories, worries, fantasies.

Then, there is ego and identity maintenance. Our thoughts often revolve around "I, me, and mine." The ego uses thought to maintain a sense of self by narrating our lives, defending beliefs, and reinforcing identity. Another reason that both psychedelics and these other long-term practices are often associated with ego-death or ego-dissolution.

Now, considering the next topic, mindfulness, let's contrast these two in order to break them down even further. In the simplest terms, meditation is actively focusing thoughts on something singular, like a mantra, a single thought, or following the voice of a guided meditation. It is not the act of shutting out all thought but rather the act of giving my mind something definite to focus on. Mindfulness involves being present with the world around you, as experienced through the five senses, filling your mind with those sensations rather than letting the DMN take over.

MINDFULNESS

Mindfulness and meditation are closely related practices, often overlapping but distinct in certain aspects. Mindfulness involves maintaining a moment-to-moment awareness of your thoughts, feelings, bodily sensations (think of the five senses), and the surrounding environment. The practice emphasizes staying in the present moment without judgment. In essence, mindfulness is a practice that can be applied at any moment throughout the day, focusing on being fully present. The goal is to observe these experiences without getting caught up in them.

Meditation is a more structured practice that often includes mindfulness but can also involve other techniques to achieve specific mental or spiritual states. Mindfulness can be seen as a subset or type of meditation; meditation encompasses a broader range of practices beyond just mindfulness.

My personal favorite way to practice mindfulness is when I am out in public. Before mindfulness, I would have been overstimulated and filled with anxiety. Now I prefer to just be a casual observer. I like to rotate through all my senses: sight, sound, smell, taste, and feel, making note of everything I can detect in each of those senses and then repeat again. I become so focused on noticing small, subtle stimuli that my mind has no place for anxiety or hypervigilance.

Practicing mindfulness has been shown to reduce stress, anxiety, and depression, improve attention and cognitive flexibility, and enhance overall well-being. It can be practiced formally (e.g., during a sitting meditation session) and informally (e.g., being mindful while eating, walking, or performing daily tasks). The goal is to develop a sustained awareness of the present moment, leading to greater self-understanding and emotional regulation.

In summary, mediation is the practice of focusing our minds on a singular thought, mantra, or guided meditation that takes us on a specific train of thought. Mindfulness in the simplest form can be sitting in the park and identifying all the sensations that your five senses are picking up. I can hear kids playing and smell someone's BBQ, I can taste the lingering taste of my lunch, I can feel the grass under my feet, and I can see the clouds shifting shape. Again, if I can maintain focus on processing the input of my five senses, I cannot simultaneously be in anxiety.

BREATHWORK

Like most of the modalities that I advocate for, it is hard to overstate the benefits of breathwork. My first breathwork session was so profound that I was immediately hooked for life. The breathwork coach guided us through a series of breathing exercises, each with its own cadence, depth, and intent. The first series was to warm up our lungs and diaphragm. Then he took us deeper. Soon after, he shifted us to box breaths while we continued to breathe and lie on our backs, then the guide led us through a meditation.

Before I knew it, I was standing in the ethereal plane, face-to-face with my younger self, the four-year-old version of me. We were gently encouraged to speak to our younger selves. I was given the gift of talking to the younger me, the gift of both asking and answering questions of myself, but more importantly, I was given the chance to reassure him/me. I wept and wept uncontrollably throughout the majority of the session. The release was intense and cleansing. I felt like I had just finished a six-hour ayahuasca ceremony.

I have since become a certified breathwork coach myself and use similar techniques in ceremony. Interestingly, breathwork can release the same DMT molecule from the pineal gland that is found in ayahuasca and can be easily achieved with practice and training.

Studies show that breathwork techniques, particularly deep and diaphragmatic breathing (breathing from your diaphragm or filling your belly), activate the parasympathetic nervous system, reducing cortisol levels and lowering heart rates. This activation helps to counteract the body's stress response, promoting relaxation and calmness. A study published in *Frontiers in Human Neuroscience* found that slow, deep breathing improved the body's stress response, particularly in those with high anxiety levels.

Techniques like coherent breathing (breathing at a rate of about five breaths per minute) have been associated with increased vagal tone, which is linked to reduced anxiety and improved emotional regulation. A randomized controlled trial published in *The Journal of Clinical Psychiatry* demonstrated that participants who practiced breathwork techniques experienced a significant reduction in anxiety and depression compared to those in the control group. Individuals practicing breathwork often report immediate feelings of calm and relaxation. For instance, many describe a noticeable reduction in physical tension and mental stress after a few minutes of deep, controlled breathing. Some practitioners describe it as a "reset button" for the mind.

Breathwork can improve cognitive functions such as attention, memory, and executive functioning. By increasing oxygen flow to the brain and reducing stress, breathwork enhances mental clarity and focus. Research in *Cognitive, Affective, & Behavioral Neuroscience* shows that controlled breathing exercises can improve attentional control and a greater ability to manage distractions.

With respect to PTSD, breathwork, particularly in the form of controlled hyperventilation (such as Holotropic Breathwork), has been used in therapeutic settings to help individuals process and release trauma. It may facilitate the emotional processing of traumatic memories by altering consciousness and allowing suppressed emotions to surface. A study in *The Journal of Trauma & Dissociation* found that breathwork sessions helped participants reduce PTSD symptoms and improve emotional regulation.

Practitioners of more intense forms of breathwork, such as Holotropic or Transformational Breathwork, often share stories of profound emotional releases during sessions. These experiences can include crying, laughter, or euphoria, which participants attribute to releasing pent-up emotions or past traumas.

Many who incorporate breathwork into their daily routines report a heightened sense of mindfulness and presence. They find themselves more attuned to their thoughts and emotions, leading to better decision-making and a greater sense of peace throughout the day. Group breathwork sessions are often described as powerful experiences that foster a sense of connection and community. Participants frequently report feeling a deep bond with others in the group, even if they are strangers, and describe the shared experience as deeply supportive and unifying.

COLD PLUNGES

You may recall that after my first ceremony, I went to Sun Valley, Idaho, and soaked in the glacier-fed lakes. That isn't the only time cold plunging has helped me regulate my central nervous system during a meltdown, as nothing else seemed to work. One of the most recent was in Sweden during my pilgrimage. As I studied with my mentor James, I was confronted with a truth that I knew but hadn't been able to resolve—my core wound, my mother wound, my abandonment wound. I had hoped to resolve it on this trip, and now I was face-to-face with the reality that the same little boy was still alone.

One of my fellow students did a journey with me. She returned from her journey, from her lower world, and related that she had discovered me. I was at the bottom of a lake, in a cabin, all by myself.

She asked me, "Where are your parents?"

"I don't have any," I replied.

This is classic Soul Retrieval, but this course wasn't about Soul Retrieval, and we were expressly forbidden from doing Soul Retrieval in this course. So, she returned and explained what she had seen. It set me off, this known reality, the fear of abandonment that had shaped my shadow—my insecurities, my low self-esteem, my fears. It had also shaped a lot of my behavior—short and frequent romantic relationships, failed marriages, and overcompensating. There it was, all of it, being related by a colleague whom I had only known for four days, but who saw it all, who had seen it all, and despite that being an integral part of my intention for traveling to Sweden, it was going to be left unresolved.

My anger, no, my rage, my frustration, my hopelessness overboiled. Soon, I found myself swimming in the swift-moving current of a Swedish river that again was being fed by glacial waters. We had been asked not to leave campus without permission and not to swim without a swim buddy, but I had blown right past both of those rules. I swim in the icy water for maybe 30 minutes until there is no more rage left inside me.

Cold plunging is one of my absolute favorite techniques for mental wellness. At home, I rarely miss a day of cold plunging, sometimes plunging two or three times a day. Cold plunging or cold-water immersion initially activates the sympathetic nervous system. However, through sustained mindfulness and breathwork, it can be navigated past the shock of the cold, leading to a switch to parasympathetic, which paradoxically can result in long-term reductions in stress and anxiety. The exposure to cold triggers a release of norepinephrine, a hormone and neurotransmitter that can elevate mood and reduce stress. Cold plunging offers an added benefit by combining three modalities—cold plunge, mindfulness, and breathwork—to distract you from the discomfort.

A study published in the *Journal of Medical Science and Sports* found that regular cold exposure led to increased norepinephrine levels, which are associated with improved mood and reduced anxiety. Cold plunging is believed to stimulate the production of endorphins, the body's natural "feel-good" and pain-killing chemicals, and can lead to improved mood and decreased symptoms of depression. Additionally, cold exposure has been linked to increased beta-endorphins and the activation of the parasympathetic nervous system, contributing to relaxation and well-being after the initial shock (approximately 30 seconds).

In my opinion, there is only one way to do it: embrace it. You have to "plunge" in, up to your chin immediately; you can't "inch" in, inch-by-inch, or your rational mind will talk you out of it. Once in, grit it out for 30 seconds. After the first 30 seconds, your body's nervous system will flood your blood with endorphins, leading to the relaxed sensations described.

A study in *Medical Hypotheses* suggests that cold-water immersion can positively affect depression symptoms due to the cold-induced release of endorphins and the shock response, which "resets" the nervous system. Regular cold plunging is often used to build mental resilience. The deliberate exposure to a stressful stimulus (cold water) teaches the body and mind to manage stress better over time. This practice can translate into greater resilience in the face of other stressors. Research in *Extreme Physiology & Medicine* shows that cold exposure can improve mental toughness and resilience by forcing the body to adapt to the stress of the cold, thereby enhancing overall stress tolerance.

Cold-water immersion has been associated with better sleep quality and faster physical and mental fatigue recovery. The cooling effect on the body can help lower core body temperature, which is conducive to falling asleep faster and enjoying deeper sleep. A *European Journal of Applied Physiology* study reported that athletes who engaged in cold-water immersion after intense training sessions experienced improved sleep quality and reduced fatigue.

Many individuals who practice cold plunging report an immediate and noticeable boost in mood. They often describe feeling invigorated and euphoric after a plunge, with some likening it to a natural high. This effect is frequently attributed to the endorphin release triggered by the cold. Regular cold plungers often share stories of heightened mental clarity and focus after their sessions. They describe feeling more alert and present, with a sharper ability to concentrate on tasks. This mental sharpness is often cited as a key reason for continuing the practice.

The act of willingly entering 50-degree water is frequently described as a powerful confidence booster. Practitioners often recount how facing and overcoming the fear of cold strengthens their resolve and carries over into other areas of life, helping them tackle challenges more confidently.

Like breathwork, cold plunging is often practiced in groups, fostering a sense of community and shared experience. Participants frequently report that the communal aspect of cold plunging, such as shared encouragement and mutual support, enhances the overall experience and helps maintain commitment to the practice.

Cold plunging combined with breathing and meditation, known as "The Wim Hof Method," has been popularized by Wim Hof. Wim is often referred to as "The Iceman" and is a Dutch extreme athlete, motivational speaker, author, and influencer known for his extraordinary ability to withstand extreme cold. He has worked with scientists to validate the effects of his methods. Clinical studies have shown that practitioners can influence their autonomic nervous system and immune response, which were previously thought to be beyond voluntary control. This research has given credibility to his claims and sparked further interest in his techniques.

Due to heavy resistance to cold plunging, I recommend trying it for 30 days before investing in equipment. You can do this in outdoor spaces, with cold showers, or by finding a friend already doing it and asking them to join daily. Even after I became convinced to add it to my daily routine, I converted a chest-style freezer into my cold plunge that I bought used for $150. It is insulated and can be cooled by plugging it in until it reaches 48 degrees.

GROUNDING OR EARTHING: THE REAL TREE HUGGERS

You may recall that I buried myself during my first ayahuasca ceremony. Subconsciously, I was drawing on past life experiences to shed years of emotional baggage. Grounding or earthing is the process of connecting our skin to Mother Earth. It is one of the easiest practices to begin. The simplest and most common grounding method is walking barefoot on natural surfaces such as grass, soil, sand, or water. This allows the body to absorb the earth's electrons and shed stray electrons, or nervous energy. You can also stand, sit, or lie directly on the ground. If standing or sitting, try maintaining contact with the earth with as much of your body as possible.

When earthing outdoors isn't an option, grounding products are available, such as earthing mats, sheets, or bands, that you can use indoors. These products are designed to connect to the earth's energy via a grounded outlet or a grounding rod placed outside. You can use these while sleeping, sitting, or working. I use a copper mat under my desk and keep my bare feet on it while I work.

Using water as a grounding tool can be effective. Standing in a natural body of water (e.g., a lake, river, or ocean) or simply placing your feet in a basin of water connected to the earth can also provide grounding benefits. And obviously, cold plunging (more beautiful overlap).

Research has shown that grounding or "earthing" can reduce stress and anxiety. For example, a study published in the *Journal of Alternative and Complementary Medicine* found that participants who practiced grounding experienced significant reductions in stress and anxiety and improved mood. This is due to the body's ability to balance its electrical state through direct contact with the earth or a grounding source.

Your body is made up of electromagnetic systems, containing an intricate and vast network of nerves and nerve endings. It is comprised of billions of nerve cells (neurons). The central nervous system alone, which includes the brain and spinal cord, has around 86 billion neurons—not to mention the peripheral nervous system, sensory nerve endings, and sensory receptors. Like any other electromagnetic machine, it requires a grounding process to protect it and offload stray or negative voltage or energy. This is the role or function of earthing.

Benefits include reduced cortisol levels, helping to stabilize the body's physiological state, calming the nervous system, and improved sleep quality. A study by Ghaly and Teplitz (2004) found that grounded participants experienced better sleep, less pain, and lower stress levels. It also synchronizes the body's internal clock and regulates circadian rhythms, leading to more restful sleep and better mental health. While direct studies on depression and grounding are limited, it has been associated with increased feelings of well-being and reduced emotional distress.

Many individuals report feeling more grounded, calm, and emotionally stable after practicing earthing. People often describe a sense of connection to nature and reduced feelings of overwhelm or stress. Practitioners frequently mention feeling more balanced and less prone to emotional highs and lows after regular grounding practice. Some people report that grounding helps improve their focus and mental clarity, particularly when feeling stressed or scattered. The practice is said to help clear the mind and bring attention back to the present moment, reducing mental distractions.

TAPPING OR EMOTIONAL FREEDOM TECHNIQUE (EFT)

My first attempt at tapping didn't go so well. A well-intentioned therapist taught me how to incorporate tapping as a way to manage my fight/flight response. Not only did I not buy into it at the time, but my wife at the time would mock me for even trying to employ it in challenging situations. Unfortunately, I quickly abandoned it. Years later, I was reacquainted with it, and again, it has become a very powerful tool in my toolbox.

Tapping is the process of tapping with your fingers on acupressure points found on your upper body. Multiple studies have demonstrated that EFT tapping can significantly reduce symptoms of anxiety and depression. A meta-analysis published in the *Journal of Nervous and Mental Disease* found that EFT was effective in reducing anxiety across various populations, with results comparable to or better than standard treatments like cognitive behavioral therapy (CBT). One of the things I love about tapping is how accessible it is to me throughout the day. I can be almost anywhere, and if I feel the need to tap, I can take a 5-minute break and address whatever it is that is coming up for me.

Tapping on specific meridian points while focusing on negative emotions or physical sensations helps regulate the body's energy system, reducing the intensity of emotional distress. EFT also holds considerable promise in treating post-traumatic stress disorder (PTSD).

A study published in the *Journal of Traumatic Stress* found that veterans who received EFT treatment experienced a significant reduction in PTSD symptoms, with effects that persisted during follow-up. EFT creates a neuroplasticity that allows for a cognitive restructuring process similar to EMDR (see below). The tapping on acupressure points and meridians creates another opportunity for neuroplasticity that enables participants to rewire the brain's response to trauma, reducing the emotional charge associated with traumatic memories.

EFT has also been used to "tap" (pun intended) into the brain's ability to control the human body. This allows participants to take control of their physical health, particularly in relation to pain management. A study published in *Explore: The Journal of Science and Healing* showed that EFT reduced pain intensity and improved overall well-being in chronic patients. By addressing the emotional components of pain, EFT may help reduce the overall perception of pain and improve the body's natural healing processes.

Many individuals who practice EFT report experiencing immediate relief from emotional distress, such as anxiety, fear, and anger. Practitioners often describe feeling a sense of calm and balance after a tapping session, even when dealing with long-standing emotional issues. Individuals implementing EFT were better able to manage their emotions in stressful situations. Many people find that regular tapping practice makes them more resilient and better able to cope with life's challenges.

People who use EFT regularly often report an overall improvement in their quality of life. This includes reduced stress levels, better sleep, and a more positive outlook on life. The sense of empowerment from having a tool to manage emotions effectively is frequently cited as a major benefit. To begin tapping, start by focusing on one specific problem, emotion, or physical sensation you want to address:

- Be clear and precise about your feelings (e.g., "I feel anxious about my upcoming presentation").
- Then, using a scale of 0 to 10, with 10 being the most intense, rate the level of distress you feel related to this issue. This helps track progress.
- Formulate a setup statement that acknowledges the issue while expressing self-acceptance. The format is usually: "Even though I [describe the problem], I deeply and completely accept myself."
- While repeating your setup statement, tap on the "Karate Chop point" on the side of your hand (either hand or both hands will do). Tap with the fingertips of your other hand, using 3-4 fingers. Repeat the setup statement three times while tapping.
- Tap on the following nine points in sequence, using 2-3 fingers to tap 5-7 times on each point. While tapping on each point, repeat a shortened reminder phrase related to your issue (e.g., "This anxiety").
  - Top of the Head: Center of the top of your head

- Eyebrow: Just above the inner corner of your eyebrow
- Side of the Eye: On the bone directly outside the corner of your eye
- Under the Eye: On the bone directly under your eye, centered
- Under the Nose: Between the bottom of your nose and the top of your upper lip
- Chin: Midway between your lower lip and the bottom of your chin
- Collarbone: Just below the collarbone, about 1 inch below where the collarbone meets the breastbone
- Under the Arm: About 4 inches below the armpit, roughly in line with a bra strap for women
- Top of the Head: Return to the top of the head

- Reassess the Intensity. After completing one round of tapping, take a deep breath and reassess the intensity of your issue on a scale of 0 to 10. Notice any changes. Repeat the tapping process if the intensity has decreased but hasn't yet reached zero.
- Repeat as Needed. Continue tapping as needed until the intensity of your issue is reduced to a level that feels manageable or reaches zero. You may need to adjust your reminder phrase as the issue evolves (e.g., "This remaining anxiety").

Example Tapping Session:

- Even though I feel stressed about work.
- Intensity Rating: 7/10.
- Setup Statement: "Even though I feel stressed about work, I deeply and completely accept myself."
- Tapping Sequence: Tap through the nine points while saying, "This stress."

Tips for Effective EFT Tapping:

- **Stay Focused:** Keep your mind on the issue you're addressing. If your thoughts wander, gently bring them back to the problem at hand.
- **Be Patient:** It's okay if you don't experience immediate relief. Sometimes, it takes multiple sessions to see significant changes.
- **Customize Your Phrases:** Feel free to adjust the setup statement and reminder phrases to better fit your specific emotions and issues.

This process can be a powerful tool for managing stress, anxiety, and other emotional challenges. Regular practice may help you gain better control over your emotional well-being.

TAI CHI AND QIGONG

I became such a big fan of Qigong that we now include it in every ceremony that we offer. On Saturdays, after an intense ceremony the night before, we needed to prepare our guests for the second night's intensity by devising a simple, easy way to help them integrate back into their bodies, and feel more grounded, before another round of medicine. We found no more effective way that was accessible to everyone, even beginners, than Qigong.

Tai Chi and Qigong have gained widespread recognition for their benefits in improving mental health. Chi or Qi both mean "life force."

Tai Chi Chuan (or Taijiquan) is believed to have originated in China during the late 16th to early 17th century. It is often attributed to Chen Wangting, a Chen village martial artist and military man, although its development is also linked to other historical figures and periods. Tai Chi evolved from traditional

Chinese martial arts and was influenced by Daoist philosophy and practices.

Qigong has a much older and broader history, with roots tracing back over 5,000 years in China. It is deeply embedded in traditional Chinese medicine, Daoist philosophy, and early Chinese spiritual practices. The practice of Qigong began with ancient Chinese practices of meditation, breathing exercises, and movement, aimed at cultivating and balancing the body's vital energy (qi). It evolved from early Daoist practices and was influenced by Confucianism and Buddhism. Qigong developed through various dynasties and periods, incorporating different techniques and systems for health, meditation, and spiritual development. It includes a wide range of practices, from simple breathing exercises to complex movement forms.

Tai Chi and Qigong are traditional Chinese practices involving mindful movement and breathing, but they have distinct characteristics and purposes.

Tai Chi originated as a martial art and includes a system of movements and forms designed to develop strength, balance, and flexibility. It often involves sequences of movements that mimic martial techniques. Tai Chi is structured around specific forms or sequences of movements. There are various styles of Tai Chi.

While modern Tai Chi is often practiced for health and relaxation, it retains elements of its martial arts origins, including principles of balance, structure, and energy flow that can be applied in self-defense and mental health. Tai Chi involves slow, flowing movements that emphasize proper posture and alignment. The practice often includes transitioning smoothly between different postures and stances. One of the forms of Tai Chi is Qigong.

Qigong is primarily focused on health, healing, and energy cultivation. It encompasses a broad range of practices designed to enhance the flow of "qi" (or "chi"), the vital life force, through the body. Qigong includes various exercises, from simple breathing techniques to complex movements. It can be practiced in standing, sitting, or lying positions.

The primary goal of Qigong is to cultivate and balance the body's energy. It often includes breath control, visualization, and gentle movement to enhance the flow of qi. Unlike Tai Chi, which is frequently practiced through specific forms or sequences, Qigong can be more flexible and individualized. It may not always follow a set sequence of movements and can be adapted to the practitioner's needs.

Both practices share the principles of mindfulness, breath control, and gentle movement, and they can complement each other well. I hope that you are seeing a pattern here. A pattern of mindfulness and breathwork is core to the others.

Practitioners often report feeling more relaxed and less stressed after regular Tai Chi or Qigong practice. This is attributed to the meditative and rhythmic nature of these exercises. Many find that the concentration required in these practices helps clear the mind and improve focus, leading to better mental clarity and reduced anxiety. The mindful movement and deep breathing associated with Tai Chi and Qigong can foster a sense of emotional stability and resilience. Participating in group classes can provide a sense of community and support, which can benefit mental health.

Research has shown that Tai Chi and Qigong can lower cortisol levels, the body's stress hormone, contributing to reduced stress and anxiety. Studies have found that these practices can enhance mood and relieve symptoms of depression. For example, Tai Chi has been shown to increase serotonin levels, which can improve mood and overall mental well-being. Regular practice can improve physical health and a greater sense of achievement, boosting self-esteem and confidence.

Research indicates that Tai Chi and Qigong can improve sleep quality, which is closely linked to better mental health. Some studies suggest that these practices may improve cognitive function, memory, mental clarity, and overall brain health. Long-term practice of Tai Chi and Qigong has been associated with increased resilience to stress, likely due to their effects on the autonomic nervous system and overall relaxation response.

YOGA

As mentioned above, with meditation, I often run into resistance when introducing some of these concepts to the masculine audience. Yoga is another practice that is met with a great deal of resistance. Thankfully, there are organizations like the Veterans Yoga Project out there that help us break down some of that resistance. Yoga is the best way I know for someone like me, someone who experiences a great deal of physical disability when it comes to joints, back issues, and pain.

Over the years, I have had three operations on just my joints, both shoulders and one ankle. This has made it difficult for me to do the activities that I spent a lifetime doing, such as running, swimming, and weightlifting. Fortunately, yoga has offered me a low-impact way to reconnect with my body, maintain fitness and flexibility, and work out. I am such a firm advocate for yoga that, looking back on my career, I am convinced that yoga should have been a core component of our Special Operations fitness regime. It would have contributed to preventative maintenance in such a way as to have prolonged our physical capabilities and prevented the level of disability we suffered.

I am sure that my resistance to yoga was similar to the same resistance I encounter now. The belief that yoga is meant for 20-30 year old women in form-fitting yoga pants and part of a community off limits to 50 year old men. The truth is that yoga is widely recognized for its mental health benefits, supported by anecdotal evidence from practitioners and empirical research. In the greater context of shamanism, the original Yogis were considered shamans.

Many yoga practitioners report feeling more relaxed and less stressed after regular practice. Combining deep breathing, mindful movement, and meditation calms the mind and body. Yoga is often described as a way to process emotions and feel more grounded. Practitioners frequently share experiences of improved emotional regulation and resilience to life's challenges. The practice of yoga encourages mindfulness—being present in the moment.

Many find that focusing on breath and body helps quiet the mind, reducing anxiety and overthinking. Many yoga practitioners claim that regular practice helps them fall asleep faster. Practices like Yoga Nidra are particularly effective for enhancing sleep quality and treating insomnia, directly impacting mental health and well-being. Practicing yoga allows individuals to develop a deeper connection with themselves, fostering a sense of self-compassion, self-acceptance, and inner peace. Attending yoga classes can provide a sense of community and belonging, which is often helpful in combating feelings of loneliness or isolation.

Yoga has been found to increase levels of gamma-aminobutyric acid (GABA), a neurotransmitter linked to mood regulation. Higher GABA levels are associated with reduced symptoms of depression and anxiety. Research suggests that yoga improves the mind-body connection, helping individuals become more attuned to their mental and physical states. This awareness helps regulate emotions and reduce stress. Regular yoga has been found to lower blood pressure and heart rate, and physiological changes associated with reduced anxiety and stress.

Emerging research suggests that yoga might enhance neuroplasticity (the brain's ability to reorganize itself) and improve cognitive functions like attention, memory, and decision-making. This is particularly helpful for individuals dealing with stress-related cognitive issues. Yoga is increasingly being used in trauma-informed care, particularly for individuals with PTSD. Certain styles of yoga (such as Hatha and Yin) can help individuals reconnect with their bodies in a safe and controlled manner, which is crucial for trauma recovery.

Keys to yoga aiding in treating mental health:

- **Breath Control (Pranayama):** Deep breathing techniques help regulate the nervous system, inducing a parasympathetic (calming) response that lowers anxiety.
- **Mindful Movement (Asanas):** Physical postures promote relaxation, body awareness, and tension release, contributing to emotional well-being.
- **Meditation and Relaxation:** Meditation practices within yoga (like mindfulness or Yoga Nidra) help quiet the mind, reduce rumination, and increase self-awareness.

The origins of yoga trace back thousands of years to ancient India, where it developed as a spiritual, philosophical, and physical discipline. Its history is rich and complex, rooted in spiritual, cultural, and philosophical traditions. But that might be more than you are after, and that's fine. Come for the flexibility and physical rehabilitative properties of yoga, stay for the well-being.

ART AND ART THERAPY

If you were to ask me, "What is the closest you can come to heaven on earth?" I'll tell you, "in the quiet of the evening, give me my art studio, a nice indica strain of cannabis, and a Fleetwood Mac playlist." This is my happiest of happy places. After a long day, or just a particularly stressful event, the fastest, easiest way for me to unwind is with a paintbrush in one hand and a joint in the other. Painting is another form of mindfulness. When I am so immersed in the creative process, my mind can't dwell on the frustrations of the day, anxiety, or uncertainty of the future.

Coincidentally, there is resistance to this as well. The resistance comes from a place of "not knowing how." The good news is that it doesn't matter. Just create. If you create something and hate it, throw it in the trash and create again. You're not trying to get an exhibit in the local art gallery; you are practicing another form of mindfulness and training your central nervous system, and trust me, your central nervous system doesn't care how your art looks. Every square inch of my garage is covered in art from my art therapy. Create your own gallery, tell your story, tell any story. Use any medium you want, use all mediums, it doesn't matter, just create something. One of my good friends that I served with runs a blacksmith's forge. This is art therapy as well.

Art therapy is a therapeutic practice that uses creative processes such as drawing, painting, and sculpture to improve mental health. Anecdotally, many people report that art therapy provides emotional relief, self-awareness, and a sense of accomplishment. Research also supports these experiences, showing that art therapy can reduce stress hormones, enhance emotional regulation, and improve mental health outcomes, particularly in cases of trauma, depression, and anxiety. Combining creative expression and psychological reflection makes art therapy a powerful tool for mental health treatment. Benefits include:

- Creating art provides a safe, non-verbal outlet for expressing difficult-to-articulate emotions. Art allows for expressing complex feelings like anger, sadness, or joy, which can be healing.
- People often report that engaging in art-making reduces stress and anxiety. The repetitive, mindful nature of drawing or painting can be calming and help shift focus away from ruminative or anxious thoughts.

- Art therapy encourages introspection. Through visual creation, individuals often gain insights into their own emotions, experiences, and inner thoughts that they might not have been able to access through talk therapy alone.
- Completing an art project gives individuals a sense of achievement. This can boost self-esteem and create a positive feedback loop of competence and emotional validation, especially for those struggling with low self-worth.
- Many participants find that art therapy equips them with new ways to cope with trauma, anxiety, and depression.
- Creating art becomes a tool they can use to self-regulate and manage emotions. For individuals who have experienced trauma, art therapy allows for a non-intrusive way to explore painful memories or events. By externalizing the trauma into an artwork, people feel a sense of control and distance, making the healing process more manageable.
- Creating art activates the brain's emotional and cognitive parts. Engaging the right hemisphere (associated with creativity and emotion) and the left hemisphere (associated with logic and language) can help process and integrate emotional experiences more effectively.
- Research indicates that the creative process of engaging in art reduces cortisol levels. This reduction is often observed even after a single session of art therapy, supporting its role in anxiety and stress relief.
- Numerous studies have shown that art therapy helps reduce symptoms of depression, particularly in people dealing with trauma, chronic illness, or grief. Art therapy fosters emotional expression, leading to catharsis and emotional processing that alleviates depressive feelings.
- Creating art can enhance neuroplasticity; this is particularly important in trauma recovery, as trauma often impacts brain structures related to memory and emotion regulation.

Art therapy helps the brain "retrain" itself to process emotions in healthier ways.

- Research has found that art therapy is another form of mindfulness—a state of present-moment awareness without judgment. Art, especially repetitive activities like coloring or drawing, can induce a flow state where individuals become immersed in the activity, leading to relaxation.
- Art therapy helps build emotional resilience by providing a structured and creative way to process emotions. Studies suggest that people who engage in regular creative activities tend to have better emotional coping skills and resilience in the face of stress.
- Studies involving older adults have shown that art therapy can improve cognitive function, memory, and emotional health, especially in those with dementia or Alzheimer's disease. The creative process stimulates brain areas responsible for cognition and emotion, helping preserve and enhance brain function.
- Group art therapy sessions can foster a sense of community and social support. Research shows that group activities can improve feelings of connection and decrease loneliness, particularly in people who struggle with social anxiety or depression.
- Art therapy bypasses the need for verbal articulation, making it particularly useful for people who find it challenging to express emotions with words.
- Creating art allows individuals to externalize internal conflicts, providing emotional relief and a sense of resolution.
- Through reflection on the artwork, individuals can gain new perspectives on their problems or emotions, leading to cognitive shifts that improve mental health.

FOREST BATHING AND REWILDING

Forest Bathing (Shinrin-Yoku)

Forest bathing is the simple act of being in the forest, allowing yourself to bathe in nature. Studies show that forest bathing reduces cortisol levels, heart rate, and blood pressure, all stress markers. Participants in forest environments often experience relaxation and reduced sympathetic nervous system activity (associated with stress). Research has found that spending time in forests decreases symptoms of anxiety, depression, and anger. Individuals also report increased feelings of vitality and well-being. Forest exposure is linked to improvements in cognitive functions like attention and memory. Nature helps reduce mental fatigue and improves focus.

If my art studio is my happiest of happy places, a close second is the forest or nature in general. It is almost impossible to be depressed in nature. Think about it objectively—how do you feel in a congested, concrete-laden, nature-less, traffic-infested, noisy, polluted area of your local big city? I know how that makes me feel. I feel like something is choking the life out of me. Now contrast that with the feelings of sitting beside a mountain lake with a gentle breeze and a blue sky. Our souls crave nature. As I write this, I am sitting in a one-room cabin in Sweden, staring out at the most pristine forest I have ever seen.

Humans have an inherent need to connect with nature, and it has a profound positive impact on mental health. Time spent in nature, whether in parks, gardens, or wilderness areas, consistently shows reductions in physiological and psychological stress markers. Being in nature encourages outdoor activities like walking or hiking, which further benefit mental health by releasing endorphins, promoting physical health, and improving sleep patterns. Research demonstrates that access to green spaces in urban areas can reduce the prevalence of mental health issues, including depression, anxiety, and social isolation.

One reason for this is that trees emit antimicrobial organic compounds called phytoncides. Breathing in phytoncides during forest bathing enhances immune function, lowers stress hormones, and improves mood. Nature demands "soft fascination" (gentle attention) rather than the "hard fascination" required by urban settings. Soft fascination provides mental rest and recovery, improving attention and cognitive clarity.

Rewilding

Rewilding is the intentional process that reconnects you with your primal traits, nature, and wildness. An event like rewilding through primal wilderness experiences involves immersing yourself in nature and tapping into your primal instincts, stripping away the distractions of modern life to connect deeply with the wild. It's about experiencing the natural world as our ancestors did, awakening our senses, and cultivating a more authentic relationship with nature. It also connects you with your more primal nature that has been separated from you due to modern conveniences. When was the last time you caught and killed your dinner, or built a shelter, or started a fire without lighter fluid or matches? It is suggested by neuroscientist Rachel Hopman that modern humans follow the "20-5-3 rule[1]."

- 20 minutes outdoors 3 times a week
- 5 hours per month in semi-wild outdoor settings
- 3 days per year spent fully immersed in nature

Spending 20 minutes outdoors several times each week has been shown to significantly reduce cortisol levels, enhance cognitive function, and foster a state of "soft fascination"—a mindful, restorative mental mode.

5 hours per month in semi-wild areas ("5"). Regular exposure—

---

1. https://www.artofmanliness.com/health-fitness/health/natures-prescription-the-20-5-3-rule-for-spending-time-outdoors/

about five hours a month—in more natural and less manicured environments (like woodlands, meadows, or preserves) helps replenish cognitive resources and deepen the mental calm that everyday routines often erode.

3 days per year fully immersed ("3"). Taking three full days per year immersed in nature (e.g., camping or wilderness retreats) can have lasting benefits. These longer, immersive experiences activate meditative brain wave patterns and can reduce stress for weeks afterward.

EMOTIONS

As I mentioned in one of my earlier journeys with ayahuasca, I saw my maternal grandfather, the WWII veteran. He had taken his PTSD and swallowed it hard, and never spoke about the war, his PTSD, or his demons. I suppose it was some sort of misdirected "that's what men do" sentiment.

That became my mother's default as well. We didn't have the difficult conversations; we didn't talk about our emotions in our home growing up. So, emotional intelligence was never my strongest suit. But to heal and reclaim our mental wellness, we have to become experts in and raise our emotional intelligence.

The good news is that there are only six basic emotions, three positive and three negative. Emotions are like a dashboard giving us real-time indicators as to the status and performance of our "vehicle." Negative emotions are like having the "check engine light" come on. They indicate an unhealed part of your soul that needs to be addressed. If you are experiencing any of the three "negative" emotions or their derivatives, you must pull over and address the check engine light.

Recent events have popularized the word "triggered." I believe that this word is misused on both sides of the debate. There are some who, when they feel triggered, feel that it is their right to hit the "easy button," putting everyone on notice that they are "triggered" and need space. The opposite side of the coin are those who want to call anyone feeling triggered "snowflakes." My shamanic view is that triggers are opportunities. Opportunities to examine why this "check engine light is blinking at me."

Continuing with the metaphor, if you are on the side that your triggers allow you to put everyone on notice and give you space, that's akin to pulling the car over and expecting traffic to halt while the car's overheated engine cools down—not very realistic. If you think everyone with a trigger is a snowflake, it is akin to ignoring your engine's warning signs and continuing to drive—also not very realistic. However, if our check engine light comes on, we can quietly pull over to the side of the road, give ourselves time to cool off (move from sympathetic to parasympathetic using these techniques), then take some time and learn why we overheated in the first place. Then, see if we can't solve that particular source of overheating on a deeper level, so as not to overheat the engine so frequently.

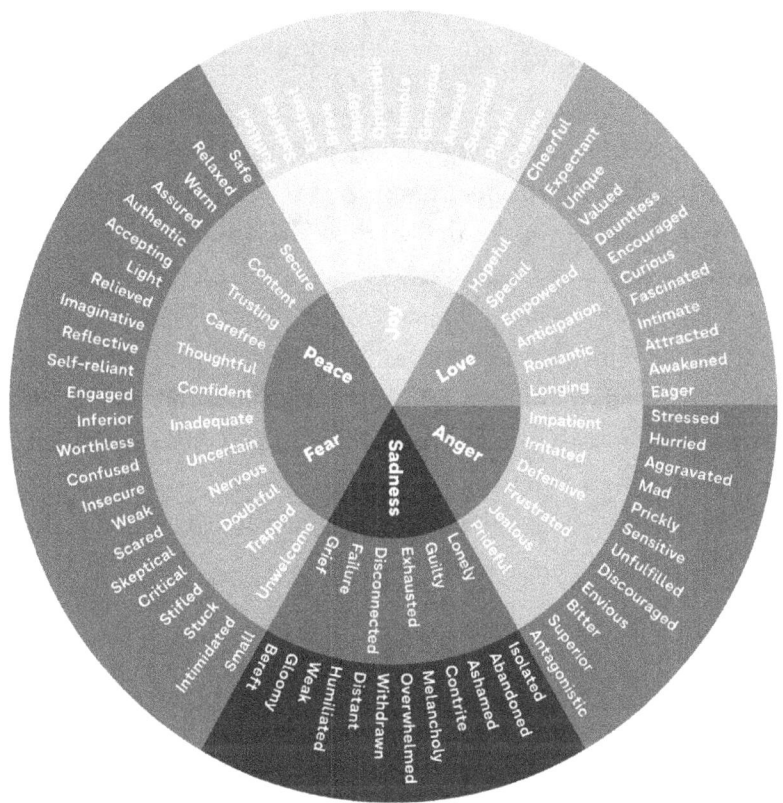

LIFESTYLE

Ignorance is bliss—a recipe.

I used to think that "ignorance is bliss" was a derogatory term aimed at ignorant people. And that the bliss they were condemned to was just ignorance. What if "ignorance is bliss" isn't a condemnation but a recipe for happiness?

I work hard at maintaining my ignorance. No news, low social media, no gossip. The less I know, the better when it comes to shallow, worthless information that our society wants to drown us in. There is very little redeeming value in it. The price you pay to be on social media far outweighs the value. Social media is to the mind what refined sugar is to the body—poison. If you want to find bliss, get ignorant.

We have reptilian brains, living with medieval systems, with space-age technology. Think about it; our ancestors never had a clue about the constant state of crisis in the world. The news outlets are not altruistic; they are not benevolent information dispensers, doing a favor for humanity. No, they are capitalistic businesses, and the business model is: if it bleeds, it leads. The juicier, the most horrific calamity of the day, becomes a headline. Your mental wellness is not their priority. I am not saying that you shouldn't stay informed or that certain causes like famine should be ignored, but I am saying that you should find reputable, impartial outlets and limit yourself to very little exposure.

THE IMPORTANCE OF THE GUT MICROBIOME

I have never been one who paid much attention to what I ate. I ate what I enjoyed. I could always control my weight as a soldier by working out, so I didn't put much thought into the food I ate. The gut microbiome was another contributor to my decline in mental health.

The happiness chemical, which fosters long-term happiness, joy, and peace, is serotonin. The gut microbiome plays a significant role in the production of serotonin. Around 95% of the body's serotonin is produced in the gut, from cells influenced by gut bacteria. The microbiome aids in making certain metabolites that regulate serotonin synthesis and signaling. This connection between gut health and serotonin levels has a direct correlation with mental health and well-being, affecting mood, anxiety, and gastrointestinal function. To maintain a healthy gut microbiome, consider the following tips:

Eat a Diverse Diet: Include a variety of fruits, vegetables, whole grains, legumes, and fermented foods.

Consume Prebiotics and Probiotics: Foods like yogurt, kefir, sauerkraut, and fibers from garlic, onions, and bananas.

Limit Processed Foods and Sugars: Reduce intake of refined sugars and artificial sweeteners.

Stay Hydrated: Drink plenty of water to support digestion.

Regular Physical Activity: Exercise helps enhance gut function.

Avoid

Processed Foods: Highly processed and refined foods can disrupt the balance of gut bacteria.

Excessive Sugar and Artificial Sweeteners: These can negatively impact gut bacteria and promote imbalance.

High-Fat Diets with Low Fiber: Low-fiber diets can reduce gut diversity.

Alcohol: Excessive alcohol consumption can harm the gut lining and microbiome.

FRIENDSHIP, CONNECTION, AND ADDICTION: FIVE CLOSET FRIENDS

Friendship significantly contributes to mental health by providing emotional support, reducing stress, and increasing feelings of belonging and self-worth. Social connections through friendship can help improve mood, combat feelings of loneliness and isolation, and offer a network of encouragement during difficult times. Also, friendships can promote positive behaviors like physical activity and healthy eating, further supporting mental well-being. Being in a circle of supportive friends also helps to build resilience and boost overall life satisfaction.

I cannot stress this enough—you need to surround yourself with good friends. Ask yourself if you are comfortable calling (insert name) at 2:00 a.m. and asking for a place to crash for a few days. Then ask yourself if you are comfortable answering the door at 2:00 a.m. and letting that person sleep on your couch for a few days. If either is a "no," then that relationship isn't on the deep level we need in our darkest hours.

The "Five Closest Friends Principle" is the idea that you are the average of the five people you spend the most time with. Attributed to motivational speaker Jim Rohn, it means that your closest relationships significantly impact your thoughts, behaviors, and overall lifestyle. These friends influence your mood, habits, and perspectives on life. Therefore, surrounding yourself with positive, supportive, and growth-oriented individuals can contribute to a healthier mindset and greater personal success. Can you name your five closest friends? Are they lifting you up or dragging you down?

FOOD, WATER, AND AIR

Everything you do—thoughts, words, and actions—affects your vibration, raising or lowering it. Everything you consume, such as food, water, air, and stimuli, also affects your vibration or frequency. If you consume processed foods, alcohol, low-quality water, and expose yourself to constant noise, air, and light pollution, you are poisoning your persona.

What we eat affects brain chemistry and mood. Processed, nutrient-poor foods can lower energy and mood, while whole, nutrient-rich foods like fruits, vegetables, and healthy fats support brain function and emotional well-being. A diet high in processed sugars can lead to mood swings and mental fog.

Hydration is crucial for mental clarity and emotional balance. Dehydration can lead to fatigue, confusion, and anxiety. Clean, pure water helps to cleanse the body, elevate energy levels, and maintain mental stability.

The quality of the air we breathe impacts mental health. Polluted air can cause brain inflammation and cognitive decline, increasing stress, anxiety, and fatigue. Fresh air and deep breathing enhance oxygen flow to the brain, raising energy levels and promoting relaxation.

Constant exposure to social media can increase stress, anxiety, and depression, especially through negative comparisons, cyberbullying, or information overload. On the other hand, mindful use of social media can raise positive connections, but should be balanced to avoid overstimulation and maintain emotional well-being. Think of everything and everyone you come in contact with or consume as something that impacts your vibration, and choose accordingly.

ALCOHOL

There is no reason for consuming alcohol. It is straight-up poison. After ayahuasca ended my alcoholism, I had a healthier relationship with alcohol. I was able to enjoy 1-2 drinks a month and never felt the constant desire to drink. But when I decided to do everything I could to raise my vibration and work on my karma, I could no longer justify those 1-2 drinks. There simply isn't any justifiable reason for anyone struggling with mental and physical health issues, or on a spiritual path, to consume alcohol. It is poison and way more dangerous than cannabis or mushrooms.

Multiple studies link alcohol, family violence, and access to weapons as significant risk factors for murder and domestic violence. Despite its extremely dangerous nature and harmfulness to the physical body, alcohol continues to be widely accepted, while plants with real medicinal value continue to be outlawed. This should tell you everything you need to know about Big Government, Big Pharma, and Big Business and their need to make you and keep you sick.

Alcohol is one of the deadliest poisons available on a mass scale in society, and yet, we don't bat much of an eye at it. It never ceases to amaze me that Veteran organizations are out there highlighting the plight of the veteran, raising money at a celebrity golf tournament sponsored by some alcoholic beverage company. This level of insanity is nearly impossible for me to understand.

With that, I want to ensure I highlight the benefits of the modalities shared.

A SUMMARY OF SHARED MENTAL HEALTH BENEFITS

Fight/Flight vs. Rest/Digest: When faced with a traumatic or stressful situation, our nervous system will turn to fight/flight for self-preservation in the short term. Over time, this can become habitual and activated even when it isn't as serious or as dangerous as the original trauma. In these less serious moments, we have the opportunity to apply techniques that halt the fight/flight response, interrupting the habit. When we incorporate one or more of these modalities, we take control of our nervous system and tell it, "Not this time, buddy, we are going to turn the ship around and head toward rest/digest."

Improved Mood and Emotional Regulation: Regular practice of these modalities enhances mood by releasing neurotransmitters like serotonin and dopamine, which can improve emotional balance and reduce symptoms of depression—the same chemicals released in a psychedelic experience. Also often overlooked is the mechanical nature of this process. If our brains have been damaged by excessive use of harmful substances like alcohol and pharmaceuticals or having TBIs, we literally have a problem with the mechanism that delivers the chemicals required to feel happy. These non-psychedelic modalities will offer some of the same neuroplasticity and neurogenesis found in psychedelics.

The Ability to be Present: These modalities improve concentration, mental clarity, and the ability to stay present in the moment. When I was suffering from TRD and suicidal ideation, I was obsessed with my past mistakes and bleak future. I could not simply enjoy the moment or be with others. I was disconnected from my present life, family, and friends. These modalities have taught me how to be present in life and have the greatest outcomes for all who begin this journey.

Enhanced Self-Awareness and Mind-Body Connection: These activities promote self-awareness, a deeper connection to the body, and improved emotional intelligence. This awareness and connection to our bodies are critical for sustained mental health. As mentioned, trauma gets trapped in our bodies. The ability to feel that trauma, locate it, and move it out of our body requires us to be conscious of our bodies. Carl Jung spoke often of "Disconnection" from our bodies as one way that we begin to experience disassociation. Dissociation magnifies our mental illness. The long-term solution is found in the modalities that follow, and they all fast-track the reunification of mind and body. The shamanic version of disassociation is "soul loss." Soul retrieval, the shamanic process of going into the lower world to find and retrieve a missing soul or soul part, is an extraordinary task and requires specific training in the process.

Better Stress Resilience: These modalities train the mind to handle stress more effectively, increasing overall resilience to challenging situations. This increases our capacity to come back from fight-or-flight or even causes us to stay in rest-digest for more extended, sustained periods, even in the face of stressful circumstances and emotional challenges. Recently, I had a conversation with a group of Green Berets, all working with meditation and breathwork. The common consensus around the table is that our use of these "follow-on" modalities, post psychedelics, has given us "objectivity" or the ability to sit as observers, with impartiality about life's ups and downs, and maintain a greater sense of neutrality.

PHYSICAL HEALTH BENEFITS

Improved Immune Function, Energy, and Vitality: Exposure to nature, cold plunging, breathwork, and Qi Gong have also been shown to boost immune response, reducing inflammation and supporting overall immunity. They also help to enhance circulation, increase energy levels, and reduce fatigue, contributing to an overall sense of vitality.

Enhanced Sleep Quality: Practices like grounding, meditation, and breathwork help improve sleep by regulating circadian rhythms, reducing anxiety, and promoting relaxation before bedtime. I found that when I was able to resolve my mental health issues and find lasting peace, my mind was able to "shut down" at night, affording me the critical rest that allowed my body to begin to heal itself.

Reduced Inflammation: Techniques such as grounding, cold plunging, and mindful practices have anti-inflammatory effects, which benefit joint health and reduce the risk of chronic diseases. Science is beginning to point to inflammation as the leading cause of chronic disease.

Better Cardiovascular Health: Breathwork, meditation, and time spent in nature can lower blood pressure, improve heart rate variability, and enhance cardiovascular health. I used to run marathons and Ultra-marathons. After I blew out my Achilles tendon in 2014, running has been impossible. Recently, my BP registered 125/76 with 63 BPMs and an oxygen saturation of 98%. Those numbers are similar to what I would have experienced as a dedicated runner, solely through breathwork and meditation.

Pain Reduction and Better Physical Recovery: Mindfulness, tapping, and meditation can assist with pain management by altering pain perception, improving relaxation, and accelerating recovery from physical exertion.

In summary: Consider a prehistoric hypothetical. I use the term *"prehistoric"* because that is when and how our central nervous systems were shaped and influenced. If you were teleported back to a prehistoric time when bears roamed the earth and were a very present danger. It is your task to go to the spring to fetch water for your family, and suddenly, a bear is ten feet in front of you. Instantly, your central nervous system switches to the sympathetic portion of the nervous system, or the fight-or-flight response. Your heart races, your breathing is shallow and rapid, your fists may clench, time is speeding up as you process possible scenarios, your heart speeds up, blood pressure increases, adrenaline and cortisol are pumping—everything in your body is rallying around surviving the next few moments.

Conversely, in the same scenario, a bear didn't appear on the trail, and you retrieved the water and returned to your family and tribe safely. You feel surrounded by safety, enjoying a meal with your family and the shelter of your cave. You are now in the parasympathetic portion of your nervous system. You eat, drink, rest, and play. Your blood pressure and heart rate decrease, you digest food, you make love to your partner, and you experience dopamine and serotonin.

The sad reality of the modern man is that most of us live constantly in scenario one, except the bears are in our heads, not stalking the water hole. The bear appears as financial shortcomings, job insecurity, shelter insecurity, global annihilation, unresolved ancestral trauma, unresolved PTSD, traffic, pollution, poisoned food, feeling of being at the mercy of Big Pharma, Big Government, Big Business. While these "modern bears" may seem much less life-threatening, in reality, our central nervous system doesn't recognize the differences. Most of the men I know live in the sympathetic, fight-or-flight mode.

An easy way to tell if you are in sympathetic or parasympathetic is to ask yourself, "I am feeling X, Y, or Z right now, and is that more likely to take place in front of the bear or back at the cave?" You can also easily determine if something is good for you and puts you into parasympathetic mode by imagining if you could do it with a bear in front of you. A modern twist is to say: "If I were practicing meditation, Cold plunging, and earthing, am I more likely to feel the danger of being on the trail with the bear, or the safety back at the cave?" Since we rarely face real bears and have the opportunity to confront the bears of our minds, we have the opportunity to reverse engineer this. If the bears of anxiety, depression, suicidal ideation, and disassociation appear on the trail in front of you, launching you into a full-blown fight-or-flight, go do the cave activities. Be present, meditate, do breathwork, take a cold plunge, have sex, enjoy a good meal, drink water, and play with your kids.

14

CORE SHAMANIC PRINCIPLES

"A shaman is not merely a medicine man or healer. A shaman is a person who journeys to other worlds in a non-ordinary reality and brings back wisdom to help and heal people." - Michael Harner, The Way of the Shaman (1980)

I get it. If you are a middle-aged male with a Christian background and rational thought, what I want to share with you will sound "crazy" at a minimum. I understand your deep resistance from experience; I was there, too. If I were reading this book in 2016, I would have dismissed this as nonsense, "woo-woo," and complete bullshit. Before this journey began, I was the biggest skeptic, an agnostic due to my religious trauma.

Most of us have been lulled asleep or put into the "Matrix" completely unaware of the mysteries and workings of the universe.

Entheogens provide the added benefit of a sincere, authentic connection to the universe. These personal interactions with what are commonly referred to as "the Divine," "Creation," "Source," or "God" are so prevalent in the entheogen space that most participants have shared experiences with the same entities. Many, like me, will experience a spiritual awakening for these reasons. I sincerely hope many of you will explore this as an intentional outcome.

In many respects, our earliest ancestors had a better understanding and relationship with the universe. They practiced an animist belief system, which attributes spiritual essence or consciousness to all things in nature, including animals, plants, rocks, rivers, and weather patterns. It holds that everything in the natural world is alive and interconnected, with spirits or souls inhabiting human and non-human entities. In many Indigenous cultures and spiritual traditions, animism forms the foundation for rituals, practices, and respect for the natural environment. It emphasizes the importance of maintaining harmony with nature and recognizing the sacredness of all forms of life.

Evidence of animistic beliefs can be found in ancient cave paintings, burial rituals, and artifacts dating back to the Paleolithic era (over 30,000 years ago). Animism still exists today in various forms, particularly within many Indigenous cultures around the world. While it may not be as prominent in modern, industrialized societies, elements of animism continue to influence spiritual and religious practices, especially in regions of Africa, Asia, Oceania, and the Americas. Additionally, contemporary movements within paganism, neo-paganism, eco-spirituality, and Shamanism often incorporate animistic principles.

My first cup of ayahuasca wiped my slate clean of any previous beliefs. Everything I ever believed was gone in one evening. This time, I had to rebuild my beliefs based on verifiable truth and within the reality of the universe I was experiencing. Don't let that scare you away from psychedelics. My experience was more extreme because I was about to step into my shamanic role as a Seidr. I have also seen all forms of belief systems in ceremonies—Catholics, Protestants, Mormons, Jews, Muslims, Buddhists, Agnostics, and Atheists. They all generally leave with the same belief system they arrived with, if not more firmly grounded in their beliefs. The largest exception is that of the agnostic or atheist.

I use the metaphor of seeing the Grand Canyon for the first time to describe what I witnessed. No amount of photos, videos, or descriptions can prepare you for the magnificence of the Grand Canyon. Standing on the southern rim, watching the sun rise is indescribable. The utter magnitude, colors, details—it is as if it is its own world, within a world with its own ecosystem, weather, and animal inhabitants. It's no wonder the Indigenous peoples recognize its sacredness.

For me, this was what stepping beyond the veil was, what seeing and experiencing the mysteries of creation and the universe felt like. Nothing can prepare you. There are no words to describe the beauty, majesty, and reality of the principles of what I will attempt to explain, but I promise it is real, more real than what most of us experience day-to-day.

The other point I wish to make is that I had a remembrance of sorts. Before ayahuasca, I wouldn't have believed any of this. After my ceremony, I simply remembered it through downloads. Wisdom, I had gathered from all my previous lives, was again available. It was as if someone had handed me a hard drive full of information that hadn't been there prior.

YOUR SUBCONSCIOUS AND THE COLLECTIVE CONSCIOUSNESS

Both are terms that are thrown around a lot in the spiritual community. Sticking with the hard drive metaphor, let me explain it like this. The pineal gland I mentioned earlier is a small, almond-sized gland, shaped like a pinecone—hence the name. It is located deep in the center of the brain. Some researchers have described it as a liquid crystal with a microscopic antenna. That's not too far-fetched when you consider that as a child, I could buy a radio assembled with crystals, wrapped in wire as the antenna and tuning mechanism. This gland, sometimes called the "third eye" in spiritual contexts, has been associated with intuition, consciousness, and various spiritual practices. This gland is like the laptop, a gateway into the non-physical world or consciousness. There are two forms of consciousness.

The first is the collective consciousness. The collective consciousness is like the internet: massive, always shifting, and available to everyone. The collective consciousness is the same; it's the massive collective knowledge of the universe. It is also always shifting as certain web domains go offline, come online, and are constantly being updated. So, too, is the collective consciousness. Every thought, word, and action every being produces is fed into the collective consciousness. And all that cumulative information, of all the universe, is available to all of us at any time. The only limiting factor is how good a laptop you have and how good your Wi-Fi connection is. Or, how good you are at using your third eye. Most of us have allowed our third eyes to atrophy. I'll go so far as to say that "the powers that be" deliberately put measures in place to separate us from our third eye or intuition.

The second form of consciousness is your subconscious. Your subconscious is like a password-protected hard drive. You and you alone have access to this information. It includes everything from your past lives to your spirit guides, soul contract, and shadow work.

CREATION

First and foremost is the creator, the divine, the universe, or, as I like to say, Father Sky and Mother Earth. Everything in this universe is made of energy. EVERYTHING. A concept supported by both science and animism. All matter—solid, liquid, gas, or plasma—is in constant motion and comprises tiny particles like atoms and subatomic components (protons, neutrons, and electrons). According to Einstein's theory of relativity ($E=mc^2$), matter and energy are interchangeable, meaning that even solid objects are forms of energy condensed into physical form.

In this view, energy permeates everything, from the largest galaxies to the smallest particles. Quantum physics further explains that particles at a subatomic level behave more like waves of energy than solid objects. This universal energy connects all things, creating an interdependent web of existence.

In spiritual and holistic traditions, this idea of energy extends to life forces like "chi", "prana," or "Od" (in Norse), suggesting that energy flows through all living beings and the natural world, affecting health, consciousness, and harmony. These traditions often seek to balance or harmonize the energy within and around us. Whether seen through science or spirituality, the universe is fundamentally an energetic system. Everything in this universe has an equal and opposite component. Father Sky is the ethereal (spirit), and Mother Earth is the material (matter) in every way, mimicking the experience of human creation. The father contributes a spark of life to the awaiting matter.

This concept of interconnectedness is vital to understand; it is the foundation of all the key elements and principles of the universe. And since all creation contains the same spark and the same matter, we are literally all connected. Not just metaphorically, or hypothetically. We are energetically all connected and all the same.

From a shamanic standpoint, all traditions adamantly discuss this concept. From the Australian Aboriginal and their Dreamtime Stories. The Native American traditions and the concept of "Mitakuye Oyasin," which means "we are all related." In Siberian Shamanism, shamans often view themselves as mediators between the human and spirit worlds, including animals, plants, and other natural elements. The Amazonian Shamanism traditions hold that plants, animals, and humans are all part of a larger web of life, and the use of sacred plants is a way to access deeper knowledge about these connections.

In Celtic Druidism, the ancient Druids of the Celtic world also had a profound respect for nature and believed in the interconnectedness of all life. Their reverence for trees, in particular, is well-documented, and they viewed certain trees as sacred beings that could connect them to the spiritual realm. These examples illustrate how various shamanic traditions have long recognized and celebrated the interconnectedness of all life, integrating this belief into their spiritual practices and worldviews. In 1935, Albert Einstein, Boris Podolsky, and Nathan Rosen published the famous EPR paper, arguing that quantum mechanics was incomplete because it implied strange nonlocal correlations between particles. Einstein later referred to this as "spooky action at a distance."

In 1982, Alain Aspect and his team conducted experiments that provided strong evidence for quantum entanglement. They used pairs of entangled photons, proving that measurements on one photon affected the other, even when separated by large distances, thus supporting this theory. In other words, what we think, say, and do impacts the universe.[1]

Timothy Leary said it this way: "The fact of the matter is that all apparent forms of matter and body are momentary clusters of energy. We are little more than flickers on a multidimensional television screen. This realization, directly experienced, can be delightful. You suddenly wake up from the delusion of separate form and hook up to the cosmic dance.

---

1. https://link.springer.com/article/10.1140/epjd/s10053-022-00542-z

TIME

You have been misled about the nature of time. Humans invented the concept of equally measured units such as seconds, minutes, hours, etc. It helps us show up at the right place and time to make life more manageable. From scientific and spiritual traditions, time is considered an illusion or a construct of human perception rather than a fundamental universal property.

In physics, especially through the lens of Einstein's theory of relativity, time is not a constant; it can vary depending on speed and gravity. This means that time is relative to the observer. For example, time moves more slowly for objects in stronger gravitational fields (like near a black hole) or for objects moving close to the speed of light. This challenges the notion of time as a universal, unchanging reality.

At the quantum level, time does not behave as we experience it. Some interpretations of quantum mechanics suggest that events at the smallest scales do not have a clear sequence of past, present, or future.[2]

From a spiritual standpoint, most, if not all, indigenous and ancient traditions support the concept that time is a human-created way of organizing reality, but in the deeper, underlying nature of existence, everything exists in an eternal present moment. In this view, the past, present, and future are all interconnected, and time is simply a tool for navigating our physical existence. It is within this shamanic belief that we begin to experience time travel, or at least a version of it.

---

2. https://link.springer.com/article/10.1007/s40509-024-00358-z

Shamanic time looks like this:

Human time looks like this:

Together, they can be experienced like this:

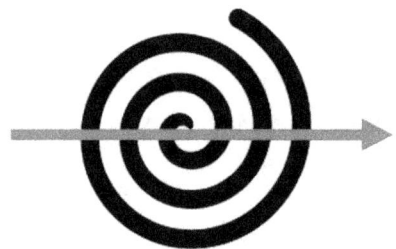

In other words, you can change the energy of past events and step into parallel universes.

THE MEDICINE WHEEL

Expanding on this concept of time is the widely held tradition of the medicine wheel. The medicine wheel is an Indigenous way of looking at the circle of life. There are multiple examples sharing similar themes, such as the balance of life, cycles of life, and interconnectedness. Examples have been found in North America in the Lakota (Sioux), Ojibwe, Blackfoot, Cree, Hopi, Diné (Navajo), Cheyenne, and Mesoamerican cultures such as the Maya and Aztec. We can see it in Andean Cultures, such as the Inca, Quechua, Aymara, Aboriginal Australians, Celtic and Druidic traditions, and the Sami (Indigenous people of Northern Europe).

While the U.S. and Canadian Indigenous peoples are most closely associated with the traditional medicine wheel, these examples show that similar concepts exist in other parts of the world, reflecting the universal human tendency to understand life through cycles, balance, and unity. When you understand the medicine wheel, you can begin to see where you are in the process. You intuitively know not to plant in winter or harvest in spring. We can time the major events of our lives with the seasons we are in, which leads us to reincarnation.

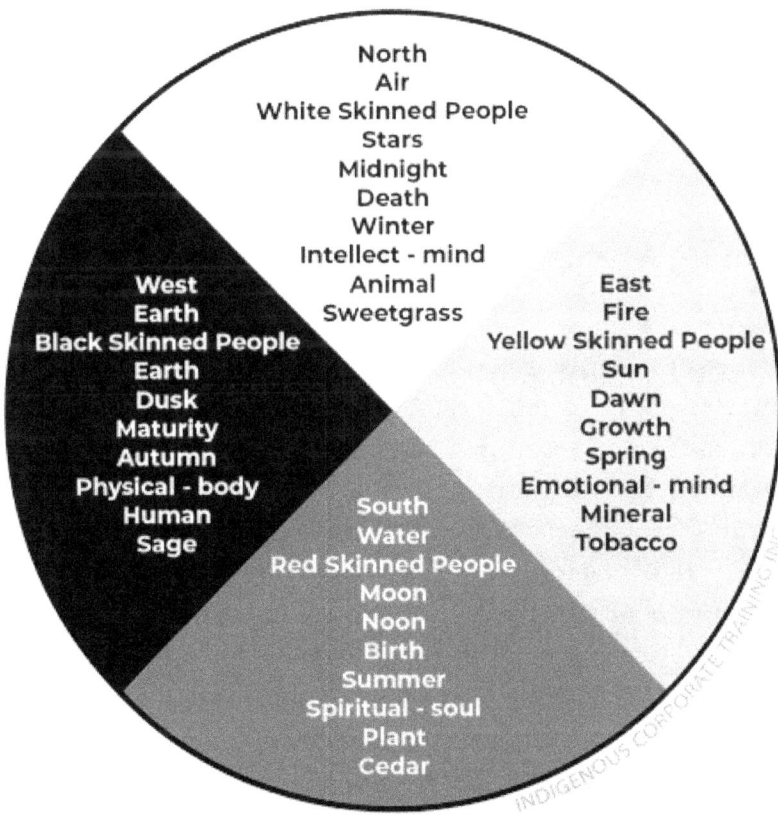

Wheel of the Year

REINCARNATION

One of the most often repeated cycles in the universe is reincarnation. Mistakenly, I once thought reincarnation was limited to only Eastern belief systems. However, many ancient and indigenous cultures around the world believe in reincarnation. Reincarnation is the widespread belief that appears across many different cultures and traditions, both ancient and modern, with slight variations in how the process and purpose of rebirth are understood. This belief is deeply tied to cycles, renewal, and spiritual growth. It is more closely associated with Hinduism and Buddhism, but found in Ancient Egypt, Druids, Aboriginal Australians, Indigenous North Americans, Indigenous African Cultures, Aztecs, Inuit, and

Greek traditions. It is especially tied to animism and Shamanism.

From a shamanic standpoint, reincarnation is the process of ascending. The process of evolving to the highest and best version of ourselves, which is called the higher self. In between each death and each birth, we continue to exist, we continue to learn and grow, work and live. Earth is one of the many realms where we can go to learn. Earth life is considered a rare opportunity and an advanced school for accelerated learning, reserved for advanced souls.

Additionally, we reincarnate as "pods" or groups. In a particular incarnation, I may be the child in the family dynamic, and in the next, my father and I might switch places. Or I may become the mother or daughter. This allows us to experience the family pod from all angles and see it from various vantage points to help us foster empathy and understanding.

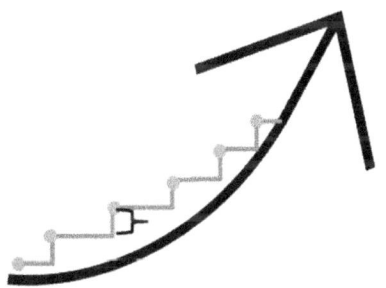

SOUL CONTRACTS

Writing our soul contract is one of the most important tasks we must complete before our next incarnation on Earth. It includes specific experiences, relationships, challenges, and lessons for spiritual growth. This agreement is made between us, the creator, the other souls who play a role in our life on Earth, and our spirit guides.

Our soul contract includes choosing our life path, including key events, people, and lessons, before birth. These experiences, often challenging, are meant to teach the soul important lessons like love, forgiveness, or compassion. Humans tend to think of contracts as "binding," but the soul contract includes free will. However, the soul contract sets certain conditions; individuals still have free will to choose how they respond and grow within the framework of the contract. The primary goal of life on Earth is to address our karmic debt.

The belief in soul contracts emphasizes that life is purposeful, with each experience contributing to a greater spiritual evolution. Evidence for the soul contract can be traced back to various ancient belief systems, although the term "soul contract" is more of a modern, spiritual phrase. They include Buddhism, Hindu philosophy, Ancient Egyptian texts, Zoroastrianism, Taoism, and Indigenous North American Beliefs. A more contemporary and Western accounting of the soul contract includes Gnosticism (Early Christian Mysticism), Celtic Druid practices, and Plato's Myth of Er (Ancient Greece). Considering Greece and Greek philosophy were the birthplace of many traditions of Western culture, why would we dismiss Ancient Greek contributions to spirituality? Most ancient belief systems rooted in animism incorporate the idea of the soul pre-selecting or being guided into specific life experiences, lessons, and challenges to facilitate spiritual growth, which aligns closely with modern interpretations of soul contracts.

Your birth is not by chance or a version of some pre-birth, Darwinian, survival-of-the-fittest sperm in a race toward the waiting egg. A fascinating new study finds that human eggs appear to "choose" which sperm will result in conceiving a baby. "Human eggs release chemicals called chemoattractants, which leave a sort of chemical breadcrumb trail that only one specific sperm uses to find the unfertilized egg," said study author John Fitzpatrick, an assistant professor in the department of zoology at Stockholm

University in Sweden. "What we didn't know until this study is that those chemical breadcrumbs act differently on different sperm, in effect choosing which sperm is successful, and misleading the other sperm away from the egg," Fitzpatrick added[3].

And Dr. Ian Scottnson, a psychiatrist at the University of Virginia, spent decades investigating children's reports of past-life memories. He documented thousands of cases where children recalled specific details about previous lives, including names, places, and events. Some of these details were later verified through investigation. His work is compiled in books like *Twenty Cases Suggestive of Reincarnation*. The principles of reincarnation and soul contracts are two of the earliest and most enduring concepts.

KARMA AND THE MEANING OF LIFE

Karma is the spiritual concept in various religions and philosophies, predominantly Hinduism and Buddhism. It refers to the principle of cause and effect, where a person's actions, whether good or bad, will influence their future experiences. Western philosophies and traditions share a similar concept, karma, though they may not use the exact term. Notable examples include Ancient Greek Philosophy, Aristotle's Concept of Virtue, Stoic Philosophy, Ancient Roman Beliefs, Gnosticism, and Pythagoreanism. These ancient Western philosophies and traditions reflect a deep engagement with ideas related to personal responsibility, moral consequence, and the impact of actions on one's spiritual journey, paralleling the concept of karma.

Karma operates on the principle that every action has consequences. Positive actions lead to favorable outcomes, while negative actions result in unfavorable experiences, with karma spanning multiple

---

3. https://www.news-medical.net/news/20200611/The-egg-decides-which-sperm-fertilizes-it.aspx

lifetimes. Karma emphasizes personal responsibility. Individuals are accountable for their actions and impacts on others, which will influence their own future experiences in this lifetime and the lifetimes to come.

There are three types of karma. Sanchita (past): Accumulated karma from past actions affects future lives. Prarabdha (present): The portion of accumulated karma being worked out in the current lifetime. Kriyamana (Future): The karma created by actions performed in the present life that will affect future experiences.

As mentioned, the goal is to transcend the cycle of karma and rebirth. This can be achieved through spiritual practices, self-realization, and living a life following moral and ethical principles. Often misunderstood, the idea that karma is seen as a form of punishment. It is more of a natural mechanism for learning and growth. It provides opportunities for individuals to correct mistakes, learn lessons, and evolve spiritually. Thus, karma guides ethical behavior and spiritual development, encouraging individuals to act with intention and mindfulness.

The meaning of life is simple—life is a school. You are here to learn, nothing more, nothing less. In fact, it is the most advanced school in the universe. More beings in the universe are waiting for their opportunity to study here at this school of advanced learning than this earthly plain can support. Viewed in that context, life on Earth is a rare and precious opportunity not to be squandered. Each incarnation is like a semester toward our ultimate goal of graduation or ascension. When we enter school, we have a long list of lessons we need to learn. The wisdom we gain and the lessons we learn and master are all carried forward from the previous semester, while those we have yet to master are still waiting for us in future incarnations.

My favorite metaphor for Karma is this: A single tree can make a million matches. And one match can burn down a million trees. That, plain and simple, is karma, my friends.

SUICIDE

I feel this is the appropriate place to discuss suicide from the shamanic perspective. My life has been shaped by suicide. Not just my long-term suicidal ideation, but also losing several close friends and tens of thousands of my fellow service members to suicide. At the time of this writing, > 6x the number that died in combat have taken their own lives. When we commit suicide (drop out) of school before the semester ends, those lessons still await us. And we still need to graduate on time, resulting in a steeper class load in subsequent reincarnations (semesters).

Suicide does not result in hell or damnation, like I was taught in the Mormon church and found in other religions. There is no shame, punishment, or judgment on the other side. And wrestling with this "demon" is part of the lesson we must master. The question is: do I want to drop out and put off these lessons for another school year, or buckle down and learn them now?

A common practice in the Veteran community is to do a "buddy check." On social media, you see the popular passive-aggressive post of "I bet I can't get ten of my friends to re-post the suicide hotline." While those are well-intentioned, I believe they fall way short of tangible concern. Instead, I suggest that the next time you speak with a close friend, follow a script that goes like this:

Me: "Hey brother, do you know I'll always pick up for you if you are in crisis?"

Them: "Yeah, I know you will. And I'll do the same for you, too."

Me: "Thanks, I appreciate that. But that's kind of easy, isn't it?"

Them: "What do you mean?"

Me: "Well, I mean that we are both in a passive role and place the burden of action on the one in crisis to call the other, right?"

Them: "Yeah."

Me: "Instead, will you promise to call me when you are in crisis?"

It is so easy to promise to pick up, especially when we know that most of us in crisis won't reach out for help. Most men carry a lifetime of conditioning that tells them it is a weakness to ask for help with anything in life, especially in our darkest times. We must flip that script and demand that our brothers pick up the phone and say, "I need help."

One time, I had a therapist tell me that if someone in crisis were to call me, the appropriate answer was to help them find professional help, essentially passing them off. I nearly had a brain aneurysm. I am my brother's keeper; that is the definition of being a brother.

Here are some best practices and practical steps to help someone who may be struggling:

- **Listen Without Judgment.** Avoid your male tendency to "fix" things immediately. Instead, allow them to talk about their feelings openly. Use phrases like "I'm here for you," or "I'm listening," to show you're present and care.
- **Ask Directly About Suicidal Thoughts.** If you suspect someone is feeling suicidal, it's okay to ask them directly: "Are you thinking about suicide?" or "Are you having thoughts of ending your life?" It shows you're willing to talk about the topic openly and doesn't "put the idea" in their head.
- **Show Empathy and Compassion .**"I'm so sorry that you're feeling this way. You don't have to go through this alone." Avoid minimizing their feelings or telling them to "just get over it." Acknowledge that their pain is real and that it's okay to feel what they're feeling.
- **Encourage Them to Seek Professional Help.** Suggest resources like therapists, counselors, or a crisis hotline. Offer

to help them find these resources or go with them if they feel comfortable. In the U.S., you can direct them to call or text the Suicide & Crisis Lifeline at **988**[4].

- **Don't Leave Them Alone (If They Are at Immediate Risk).** Stay with them until help arrives, or you can connect them with professional support. Call emergency services for assistance, even if the person is reluctant. Their safety is the most important concern.
- **Help Them Create a Safety Plan.** Help them develop a safety plan when overwhelmed. This plan might include:
  - A list of supportive people they can contact (friends, family, therapists).
  - Coping strategies that have worked for them in the past.
  - Activities that bring comfort and grounding.
- **Check in Regularly.** Not just in moments of crisis. Let them know you care consistently. Messages like "I'm thinking of you" or "How are you doing today?" can help them feel supported and less isolated.
- **Know Your Limits and Seek Support.** Make sure you're also taking care of your mental health. Seek advice or support from mental health professionals on how to best support your friend or loved one.
- **Educate Yourself.** Understanding more about suicide and mental health can help you better understand what your friend is going through. Resources like the National Alliance on Mental Illness (NAMI) or the American Foundation for Suicide Prevention (AFSP) can offer more information and support.

If you or someone else is experiencing a crisis, please contact a professional or call a local emergency number immediately. **The suicide hotline is 9-8-8**

---

4. I have used the suicide hotline myself in my times of crisis.

KARMA

Balancing karma in this lifetime involves taking conscious steps to resolve past actions and cultivate positive energy. While different traditions may offer specific rituals or methods, here are some common principles and practices for clearing karma:

- Self-Awareness and Accountability
  - Recognize Past Actions: The first step to clearing karma is acknowledging any negative actions, thoughts, or behaviors from this life or past lives. Reflect on patterns that may have caused harm to yourself or others.
  - Take Responsibility: Accept responsibility for the consequences of your actions. This process requires emotional honesty and humility to recognize where you've caused pain or imbalance.
- Forgiveness
  - Forgive Others: Holding grudges or resentment creates negative karmic energy. Practice forgiveness, not as a way of condoning harmful behavior, but as a means of freeing yourself from the emotional and energetic burdens of the past.
  - Self-Forgiveness: Equally important is forgiving yourself for mistakes and actions that may have caused harm. Self-compassion allows you to move forward without guilt, which can be a heavy karmic burden.
- Resolve Conflicts and Make Amends
  - Apologize and Make Amends: Healing karmic relationships can often involve reaching out to those you may have wronged, offering a sincere apology, and making reparations where possible.
  - Reconcile with Yourself: It's also important to resolve internal conflicts, such as feelings of guilt, shame, or fear, as these can perpetuate negative karmic cycles.

- Compassion and Service to Others
  - Acts of Kindness: Positive actions toward others generate good karma. Helping those in need, practicing empathy, and showing compassion are powerful ways to balance or clear karmic debts.
  - Service (Seva): Many spiritual traditions emphasize selfless service or seva to neutralize karma. Volunteering, helping the community, or performing acts of generosity without expecting anything in return can shift your karmic energy.
- Meditation and Mindfulness
  - Meditation: Practices like meditation, particularly those focused on compassion (e.g., Loving-Kindness Meditation) or inner awareness, can help clear emotional and mental patterns tied to negative karma. Meditation fosters awareness of harmful tendencies and helps dissolve them.
  - Mindfulness: Being present and mindful in daily actions helps prevent the creation of new negative karma. It allows you to act from a place of awareness rather than reacting unconsciously to situations.
- Spiritual Practice and Prayer
  - Prayer: In various traditions, prayers of repentance or intention to heal karma are common. This might involve asking for guidance from higher powers, such as God, deities, or ancestors, to help clear karma and set the intention for healing.
  - Mantras and Rituals: In Hinduism and Buddhism, certain mantras (e.g., the Mahamrityunjaya Mantra in Hinduism or Om Mani Padme Hum in Buddhism) are believed to help clear negative karma. Performing rituals, such as lighting candles or offering prayers, can also symbolize releasing past burdens.
- Healing Emotional and Energetic Blockages

- Energy Healing: Practices like Reiki, Qi Gong, or Pranic Healing are thought to cleanse negative energy accumulated from past karma. These healing practices balance your chakras or energy centers, helping clear karmic patterns.
- Shamanic Practices: Some people seek shamanic healing to remove karmic blockages. Shamans often work with spiritual guides or energies to identify and heal karmic imbalances.
- Living Dharma
  - Live According to Spiritual Principles: In Hinduism and Buddhism, living in accordance with dharma (righteousness, duty, and truth) helps to counteract negative karma. This means aligning your actions with universal principles of compassion, nonviolence, and integrity.
  - Right Action: In Buddhism, the Noble Eightfold Path offers guidance on how to live ethically and wisely. Practicing "right action" and "right speech" helps to reduce harmful behaviors that generate negative karma.
- Conscious Intentions and Choices
  - Set Positive Intentions: Every action begins with intention. Set positive intentions to create good karma and approach life with love, kindness, and humility.
  - Break Negative Cycles: Notice any negative behavior patterns or recurring difficulties. Consciously choose to respond differently to break these karmic cycles.
- Karmic Healing with Professionals
  - Past-Life Regression Therapy: Some people seek past-life regression therapy to uncover unresolved karma from previous lifetimes. This process helps individuals identify karmic imprints and work through them consciously.

- Spiritual Counseling: Working with a spiritual teacher or counselor can help you gain insight into karmic patterns and provide guidance on how to transform negative karma.
- Living in Alignment with Nature and the Universe
  - Harmony with Nature: Many traditions believe that living in balance with nature and respecting the environment can clear negative karma. This involves practicing sustainability, respecting all living beings, and living harmoniously.
  - Gratitude: Cultivating a daily practice of gratitude shifts your energy and consciousness toward abundance and positivity, reducing the impact of past karmic negativity.

You can clear and resolve karmic energy in this lifetime by becoming more conscious of your actions and consequences, practicing kindness, making amends, and aligning your life with higher spiritual values. Addressing our karma is how we deal with ego.

EGO

Sigmund Freud popularized the term ego. The ego is the part of the mind that mediates between the id (the primal, unconscious desires) and the superego (the internalized moral standards). Freud saw the ego as the "executive" of the personality, responsible for decision-making, reality testing, and balancing the demands of the id, superego, and external world.

The ego can contribute to our karmic debt in the actions, thoughts, and behaviors, often rooted in selfishness, ignorance, or harm caused to others. The ego is frequently focused on personal desires, needs, and ambitions. When it leads to actions that harm others or are motivated by selfishness, these actions can create negative karma. In spiritual traditions like Buddhism and Hinduism, the ego fosters attachment to material things, status, or personal identity. These attachments can lead to desires and actions that perpetuate suffering for oneself and others, adding to karmic debt. The ego reinforces the idea that we are separate from others and the universe. This illusion can cause actions based on fear, competition, or aggression, which create negative consequences that add to karma. The ego often leads to judgment, resentment, and conflict, resulting in harmful actions. When driven by ego, these actions can cause harm and suffering, generating karmic debt.

Most people in the psychedelic and spiritual space discuss the "killing" of our egos. This is nonsense. You cannot kill off part of your being; it is a major part or aspect of every person and serves a divine purpose. Our egos play a sacred role as well. They often keep us safe, sometimes causing temporary or permanent disassociation from the traumatic experiences that we are experiencing (more on that in a moment). I also believe that we cannot kill our egos, but we can tame them. We say, "thanks for keeping me safe. I will take over now; please move over to the passenger seat." Shamanically, this disassociation is called soul loss.

SOUL LOSS AND SOUL RETRIEVAL

Soul loss is the shamanic term for disassociation, a mental process where a person disconnects from their thoughts, feelings, memories, sense of identity, and, in extreme cases, even the sensation of being connected to their own body. It is often a coping mechanism in response to trauma or stress, allowing the person to distance themselves from painful experiences.

Dissociation can manifest in various forms, from mild daydreaming to more severe conditions like Dissociative Identity Disorder (DID), where a person may have multiple distinct identities or personality states. In psychology, dissociation is a protective response that helps individuals manage overwhelming experiences by separating the traumatic event from their conscious awareness. Jung, the colleague and student of Freud, and the father of modern psychiatric treatment, recognized that trauma, neglect, or repression could lead to fragmentation within the psyche, where parts of the self become split off or disconnected from conscious awareness. This fragmentation is akin to what might be described as "soul loss" in shamanic terms. Reintegration: In psychological treatment, reintegration often involves therapy aimed at helping individuals reconnect with their dissociated parts, bringing them into conscious awareness and working through the trauma. Similarly, in soul retrieval, the shaman reintegrates the lost soul parts to restore balance and well-being.

The concept of soul retrieval is deeply rooted in shamanic traditions worldwide, with some of the earliest documented instances of practices resembling soul retrieval in various ancient cultures.

Shamans practiced soul retrieval for thousands of years among the Indigenous peoples of Siberia. Early anthropologists and explorers documented these practices, but they predate written records by millennia. While not directly referring to soul retrieval, the ancient Egyptians had a complex understanding of the soul, consisting of multiple parts (such as the ka, ba, and akh). Some of their rituals and texts, like the Pyramid Texts and the Book of the Dead, involve practices to ensure the soul's safe passage and reintegration in the afterlife, which can be seen as analogous to the concept of soul retrieval. Greek mythology and literature contain references to the journey of the soul and the afterlife, such as the myth of Orpheus and Eurydice, where Orpheus ventures into the Underworld to retrieve his wife's soul. While this is more mythological, it reflects a cultural awareness of the soul's journey and the possibility of retrieval.

In the 19th and 20th centuries, anthropologists like Mircea Eliade, Carlos Castaneda, and Michael Harner studied and documented shamanic practices, including soul retrieval, among various

Indigenous cultures. Through his book *The Way of the Shaman*, Harner's work has brought widespread attention to the practice of soul retrieval in contemporary times. Dissociation in psychology and shamanic soul retrieval describe different frameworks for understanding and addressing the disconnection within a person; they share a common goal of reintegration and healing. Both approaches recognize that trauma can lead to a loss of self and emphasize the importance of bringing these fragmented parts back together to achieve a sense of wholeness.

SHADOW WORK

Shadow work is a concept introduced by Carl Jung that involves exploring and integrating the unconscious aspects of our personality, known as the "shadow." Shadow work is about confronting and embracing the hidden, often uncomfortable aspects of the self to achieve greater self-awareness and wholeness. By integrating the shadow, individuals can lead more balanced, authentic lives.

The shadow consists of traits, desires, and emotions we repress or deny because they don't fit our ideal self-image or societal expectations. These can include fears, insecurities, or even hidden strengths. According to Jung, the shadow represents the parts of ourselves we find undesirable or unacceptable and therefore suppress. It can include negative traits like anger or jealousy, and positive traits like creativity or assertiveness that we've been taught to suppress. Shadow work is the process of bringing these hidden parts into conscious awareness. Acknowledging and accepting them, we integrate the shadow into our overall personality, leading to personal growth and wholeness. Shadow work helps reduce inner conflict, increases self-awareness, and allows for more authentic self-expression. It also improves relationships by reducing projections (when we see what we deny ourselves in others).

Jung created the perfect term with "shadow" because those undesirable traits are always with us; there is no way to separate ourselves from them. We can extend this metaphor and ask ourselves, where is the "shadow" the most and the least prominent? The most prominent would be far from the equator, casting an extremely long shadow at sunrise or sunset. The least prominent would be on the equator, on the summer solstice, at high noon, where it would be almost indistinguishable. We must drag our shadow components out into the light of day and own them in truth and honesty for them to no longer have control over us.

SPIRIT GUIDES

Soul contracts, ego, and shadow work can all begin to feel overwhelming, but a whole team is surrounding you to support you. Spirit guides are non-physical beings committed to assisting you during your time on Earth. They signed off on your soul contract with you. Your spirit guides have been assigned to you and involved in your life since birth, providing guidance, protection, and support. You may have experienced them through "signs," coincidences, synchronicities, dreams, and symbols. They are the ones behind the "breadcrumbs." They can also serve to protect you.

Spirit guides are an ancient concept originating from animism and are found in all ancient spiritual practices worldwide. These traditions often have their own unique interpretations of spirit guides, but they share the common belief that non-physical entities provide guidance, wisdom, and protection. Spirit guides appear in many ancient traditions worldwide, taking the form of animal spirits, ancestors, gods, angels, and nature spirits. These guides offer wisdom, protection, and guidance, helping individuals navigate life's challenges and maintain harmony with the spiritual world. Each tradition incorporates unique practices for connecting with these guides, from shamanic rituals to ancestor veneration and meditation.

Here are some ancient traditions that incorporate the concept of spirit guides: Native American Spirituality, Ancient Egyptian Religion, Celtic Druidism, Hinduism, Buddhism, Ancient Greek Religion, African Traditional Religions, Chinese Spirituality (Taoism and Confucianism), Aboriginal Australian Spirituality, Christianity and Judaism, and Kabbalah.

Types of Spirit Guides:

- Ancestors: Persons in your ancestral lineage several generations past. They offer protection and guidance based on shared lineage. After all, you can play a key role in clearing their karma.
- Angels: Spiritual beings, often associated with light and love, are believed to protect and offer divine guidance. This includes Archangels and Guardian angels. Your guardian angel has a specific role—keeping you from a premature death, before completing your soul contract. A person may have more than one guardian angel; I believe I have a few. But a guardian angel will only be assigned to a single person.
- Animal Spirits: Often called totem animals or power animals, they represent specific qualities and strengths that can help guide a person through life.
- Ascended Masters: Enlightened beings such as Buddha, Jesus, or other spiritually evolved figures who offer wisdom and higher knowledge.
- Elemental Guides: Spirits connected to nature and the elements (earth, water, fire, air) that guide individuals toward harmony with the natural world.

Roles of Spirit Guides

Spirit guides are believed to help individuals make decisions, avoid harm, and stay aligned with their spiritual path. They provide intuitive nudges or signs to help people navigate tricky situations or understand more profound truths. Spirit guides are thought to assist in emotional, spiritual, and physical healing, offering comfort and strength during challenging times.

They guide individuals through life lessons, helping them grow spiritually and fulfill their purpose. Many people connect with spirit guides through quiet reflection, meditation, or prayer, opening their minds to intuitive messages. Spirit guides often communicate through signs (like repeating numbers or symbols) or through dreams and synchronicities. Some tools, like tarot, pendulums, or automatic writing, are used to receive direct guidance from spirit guides.

RELIGION, SPIRITUALITY, AND CONSCIOUSNESS

I want to be very clear about this. I do not have an axe to grind against religion or even the Mormon religion. Religion is not wholly virtuous, as ardent followers would have you believe. It is not wholly evil, as those opposed to religion would have you believe, either. Its virtue or harm is more likely determined case by case, and examples of both are found in each institution. One thing can be said—all religions are man-made interpretations of spirituality to explain creation and prescribe dogma for how to engage with said creation.

Don't let this next comment trigger you; it is meant to be a metaphor, not an insult. But religion is like a "crutch." If you have a broken leg and need support, religion can be great; it can offer you the support, social structure, and moral compass you need. However, once your leg has healed and you are ready to run, the crutch slows you down. When prepared to set aside religion, you can step into spirituality.

Spirituality is the individual, intimate process of seeking connection with a higher power, presence, or universal truth. It is unencumbered by religious doctrine, offering individuals a path to experience the sacred in a more authentic, personal, and meaningful way.

Individuals often experience transcendence, awe, or connection, guided by an inner knowing or higher consciousness. But in truth, spirituality is just another crutch as well; the goal is consciousness or transcendence.

Consciousness represents a more expansive and transcendent state of awareness beyond the search for meaning or connection with the divine. It is the direct experience of being, where the boundaries between self, others, the divine, and the material world dissolve into a unified field of existence. Consciousness is not just a way to perceive or understand reality; it *is* reality. It moves beyond the duality of seeking and finding to where the mind quiets and the experience of self as an individual ego dissolves.

Consciousness is the highest level of human experience—a state of being where one is fully present in the eternal now, without attachment to past or future, without the need for labels, beliefs, or external validation. It is the oneness with all that is, transcending the mind, emotions, and even spiritual constructs. It is the pure awareness that everything arises from the same source, and that source is within and beyond you simultaneously. It is being fully immersed in the essence of existence itself, where even the idea of a "path" to something greater dissolves into the recognition that there is only now, only here, and only this boundless awareness.

Transcendence	I am God (as are you)
Spirituality	I have my own relationship with God
Religion	There is a God and I prefer to allow someone else to intermediate for me and God
Agnosticism	There may or may not be a God
Atheism	There is no God

LETTING GO OF ATTACHMENT, RADICAL NEUTRALITY, AND EQUALITY

One of my favorite philosophies is letting go of attachment. Simply put, our attachment to people, things, desires, or outcomes causes suffering. This concept is rooted in many spiritual traditions, which teach that attachment leads to dissatisfaction because everything in life is impanent. By releasing attachment, individuals find freedom, peace, and greater clarity.

Impermanence: Everything changes, so holding tightly to anything is futile. Accepting change allows for inner peace. People often think they "own" things, relationships, or identities, but everything is borrowed, including time and experiences. Letting go requires surrendering control over outcomes and trusting in the flow of life. By letting go of attachment to the past or future (concepts that don't exist), one can fully live in the present moment, appreciating life as it is. Detachment brings a sense of freedom, allowing individuals to live authentically without being weighed down by expectations, possessions, or fears. It encourages embracing life with openness and accepting things as they come and go.

Which brings me to radical neutrality. I try to practice radical neutrality in every aspect of my personal life, having lived extremism in my own religious experience and seeing up close and personal the effects of radical Islam. According to universal law, evident by nature, balance will always be the true end goal. Sure, the lion is at the top of the food chain . . . until it's not. Until it dies and is consumed by a fly or worm. Extremism is where we get into danger. It doesn't matter if we are talking about politics, religion, diet, etc. Anything taken to its extreme harms our balance and centeredness.

The ascension or achieving consciousness is rooted in the concept that "we are all equal." Everyone who has ever walked or will walk this earth is your equal. And it isn't limited to humans. It includes everything on this planet—the rocks, animals, trees, and insects. This is why the followers of Buddha wouldn't harm a fly, the idea that the fly could likely be your reincarnated grandmother.

Again, it is our ego at work. Most of us falsely believe that ego is when we have an exaggerated sense of superiority, when in fact our ego is distorting our self-view of equality in both directions. In spiritual and psychological contexts, the ego refers to the sense of self or identity that separates us from others. Both superiority and inferiority are rooted in the false identification with a limited sense of self. In both cases, the ego clings to a limited sense of self defined by external factors like achievements, failures, possessions, or others' opinions. The true self is not defined by comparison, but by unity, compassion, and a broader understanding of self and others.

In my egoic state, I would constantly judge others. Everything was a competition. If I felt someone was better than me, I'd find ways to find their flaws. When I thought I was better than someone, I enjoyed the false sense of superiority. One of the ways I overcame my tendency to judge others is simply to remind myself that we are all connected and all equal. No single person on earth is better than or worse than me. This can be a humbling experience to see the homeless, the criminal, the addict as equals, just living out a different soul contract, and in a different part of the journey than I am. They just happen to be enrolled in other courses this semester, that's all.

BALANCED AND CENTERED

If you recall, this was one of the earliest lessons that Je taught me. It means that we have to be balanced in our masculine and feminine energies, bringing all four of them into play, becoming the whole being, not just living from a place representing this incarnation's embodiment. In other words, just because you may have been born as a male, don't lose touch with your emotions or spirituality; this is balance, in the horizontal plane. Your masculine energy is on your right side, and your feminine energy is on your left.

Centeredness is striving for connection above and below. In spiritual and religious communities, we often hear about the importance of being connected to a "higher" source. What about "source energy" below our feet? Father Sky, the center of the universe, is both a physical and spiritual source of energy. The center of Mother Earth is also a divine source of energy, both physically and spiritually. We have been taught all our lives that our spirit inhabits our bodies. This is not true. Our bodies inhabit our spirit. Our spirits extend beyond our bodies in what is known as an aura or auric field. Both the body and the spirit make up the whole persona, and both are conduits of energy. Our personas need to be connected in such a way that allows energy to flow through us. It is just as harmful to constantly strive for the source energy above us while neglecting the source energy below us, and vice versa.

We achieve stability by balancing all four aspects of your personas, two masculine and two feminine, and strength by being connected to both of our source energies.

The right side of our persona is masculine (Mental and Physical attributes)

When we are "balanced" all four attributes of our persona are functioning optimally.
We are centered when we are both rooted in the physical realm and stretching for the astral realm.

The left side of our persona is feminine (Spiritual and Emotional attributes)

HO'OPONOPONO, THE PRAYER OF FORGIVENESS

My first experience with Ho'oponopono was so powerful that I share it with everyone I can. I was struggling with a personal relationship. I had become estranged from someone who was very important to me. When I learned about this prayer, every time I thought of this person or felt the knife-like stab in my heart over the loss of this relationship, I said this prayer. Over the course of a year, the relationship was mended, and this person came back into my life.

Ho'oponopono was popularized in mainstream spirituality by Dr. Ihaleakala Hew Len, a Hawaiian psychologist, and his teacher, Morrnah Nalamaku Simeona, a kahuna lapa'au (Hawaiian shaman). Morrnah modernized the traditional Ho'oponopono practice in the 1980s, adapting it from a group process to an individual practice. She introduced the idea that one could heal the world around them by healing oneself. However, Dr. Hew Len's story truly captured the mainstream's attention. Dr. Len was hired as a staff psychologist at the Hawaii State Hospital in the 1980s, specifically in a ward for criminally insane patients. These patients were considered the most dangerous, violent, and unmanageable. With nothing more than a practice of Ho'oponopono, Dr. Len healed an entire ward of mentally ill criminal patients without ever seeing them in person, by repeating Ho'oponopono and focusing on healing his responsibility for the patients' conditions, and the perceptions and judgments of the patients. His approach emphasized the idea of 100% personal responsibility for everything one experiences.

Now, sit with that for a moment. He prayed, taking personal responsibility for the crimes of inmates he never knew or met. Is there any logical reason that he was responsible for the crimes of these violent, mentally ill persons in his care? No. No, there wasn't. But spiritually speaking, we are all connected, we are all enrolled in the same semester at the same university, and we are all working on our individual and collective karma. In the spiritual sense, we are all an extension of the same source energy, all connected and responsible for each other. In that context, we are all accountable for the forgiveness process regardless of whether we are the villain or the victim. That's deep philosophy, but true. Ho'oponopono promotes inner peace, self-purification, and healing by clearing away emotional blocks and fostering love and forgiveness.

The practice is simple and revolves around repeating four key phrases. Here is a simple version:

- I'm sorry—Acknowledging personal responsibility for any disharmony, whether conscious or unconscious.
- Please forgive me—Asking for forgiveness for any harm caused.
- Thank you—Expressing gratitude for the opportunity to heal and for the forgiveness received.
- I love you—Affirming love, which is seen as the ultimate healing force.

Here is a more comprehensive version, similar to the one I mentioned above:

- I'm sorry for causing you pain.
- Please forgive me for not showing up for you the way I should have.
- Thank you for being a light in my life.
- I love you forever and unconditionally.

THE DIVINE MASCULINE

"The more the man becomes conscious of his own inner woman, the more he will understand the dynamics of his own masculinity. In the same way, the more the woman becomes conscious of her own inner man, the more she will understand her own femininity." —Carl Jung.

There has been much said about masculinity today, resulting in various descriptions-

- Toxic masculinity: the theory that somehow masculinity is poisonous.
- The Crisis of Masculinity: the idea that traditional roles and values are being undermined.
- Simps: men who embrace submissiveness to please their partner.
- Hypermasculinity: Refers to exaggerated male traits, such as extreme aggression, physical strength, and sexual dominance, often seen as a reaction to societal changes.

It is obvious that there has never been a time in history that has experienced such diverse and varied sources of influence, causing pressure on the role of masculinity. Several societal factors continue to contribute to the issues, particularly concerns about men feeling a loss of identity or purpose. These influences can shape how masculinity is perceived, both in a traditional and modern sense.

The Shamanic reality is this. You have no gender! Take a moment, go get a drink of water if this is upsetting. You are not a man; you are not a woman. You are an infinite being of light, a fractal of God, made up of energy from the stars[5], inhabiting bodies made from recycled elements from dinosaurs, animals, plants, and butterfly wings, riding a blue crystal through an ever-expanding universe.

---

5. All Iron, as found in your red blood cells comes from the death of massive stars and released into the universe when those stars die in a supernova, seeding the cosmos with the building blocks of life. This is the only known source of iron in the Universe. In other words you have stardust in your veins.

As infinite beings of light, you have four parts to your persona.

- Masculine
  - The physical body
  - The mental, or knowledge, you possess
- Feminine
  - The spiritual energy
  - The emotional

There is a fine line with respect to our bodies. On one hand, some live with complete disregard for their bodies; on the other, some place a high degree of emphasis on beauty and physical perfection-both are extreme, and both are dangerous. The Shamanic view of your body is important. You chose your gender, your sexual orientation, your race, and all your physical features as part of this incarnation only. For specific reasons, the lessons you need to learn, characteristics that you need to master, pertain to maintaining your body. The physical body makes up one-half of your persona; it is important. However, it isn't so important that you prioritize your body over your spiritual growth. When you consider that you have reincarnated hundreds of times, it can help put into perspective just how temporary physical beauty can be. It is important to treat your body with respect and care for it in a way that allows you to live your fullest life, for the purposes you chose to come into the world.

To realize our full potential during this and every incarnation, all of us must learn to bring all four elements into harmony and balance. This is what it means to be in your divine masculine. Anything less than harmony and balance robs us of our full potential. I like to use the metaphor of a one-legged fighter in the ring. I am pretty sure I could beat the greatest fighter of all time, Muhammad Ali, if he had one arm and one leg tied behind his back. If he had to chase me around the ring by hopping, if I could knock him over with a push or a shove. There would be no "dancing like a butterfly" and no "stinging like a bee." Having both legs and both feet is what creates balance, flow, energy, and power. As a human, if you choose to ignore two key components of your being (spiritual and emotional), you are no more effective than a one-legged boxer, no matter how much protein you eat, how much you lift, how driven you are, how much worldly success you have—you are still lacking. I am speaking from experience here. I was that guy, and it left me sick, unbalanced, and, worst of all, chronically unhappy.

It doesn't help that in our contemporary Western society, we have also abandoned the ceremonial rites of passage for boys transitioning to manhood. These ceremonies often involved physical, emotional, and spiritual challenges to prepare boys for adult responsibilities within the tribe. Ironically, the term "hero dose" in the psychedelic context refers to classical literature and Greek myths of initiation. In both classical literature and Greek myths, the boy embarks on a transformative journey where the hero undergoes trials, gains wisdom, and returns transformed. Psychedelic experiences can evoke similar journeys, leading individuals through introspection and profound insights about themselves and their place in the universe or tribe. There are also plenty of non-psychedelic ceremonial ways to assist boys in the process of becoming men, such as vision quests, pilgrimages, and others. Without this formal process, young males often fail to find and maintain healthy masculinity.

Aboriginal tribes in Australia practiced long periods of isolation in the wilderness, where boys learned survival skills, tribal laws, and sacred knowledge from elder men. They returned as men upon completion, ready to take on adult roles. Maasai boys underwent an initiation, spending months to years living away from the community, learning warrior skills, such as hunting. Upon their return, they were considered full warriors (Moran) and men within the tribe.

Native American tribes, like the Apache, had vision quests or spiritual journeys for boys transitioning to manhood. Boys would spend time in isolation, fasting, and praying, seeking guidance from the spirit world for their path in life. Upon returning with a vision or spiritual insight, they were recognized as men by the tribe, to name a few.

Don't get me wrong. For all the criticism I have of the government, war, and the failures to appropriately address the needs of our veterans, the warrior is a sacred role and archetype. Four main and commonly held masculine archetypes are the King, Lover, Magician, and Warrior. The King is the most important archetype, incorporating the other three archetypes. It's often the last archetype to develop in a man's life. The Magician is associated with turning setbacks into opportunities for growth. Magicians are mediators between the human world and the Divine world, and they can explain spiritual ideas in ways that others can understand. The Lover is associated with passion, sex, and living life fully. It's often the first archetype to develop in a man and is associated with youthful idealism. The Warrior is associated with destruction, but creative destruction makes room for something new. Warriors are emotionally detached when on a mission and need a single-minded purpose to achieve it.

Psychologist Robert Moore and mythologist Douglas Gillette identified these archetypes in their 1990 book, King, Warrior, Magician, Lover. Carl Jung initially developed the archetypes. There is a sacred and primarily masculine responsibility for the defense of society. The Neolithic period saw us move from hunter-gatherers to farmers with fixed villages. These villages provided rich targets for those robbing and pillaging these communities. One of the roles of the divine masculine was to provide a boundary or wall around these settlements for the divine feminine to flourish, nurture, and reproduce. There is a sacred aspect to being a soldier and fulfilling this archetype. Where it goes sideways is with the corruption of politicians and the military industrial complex, turning the sacred act of defense into a power and money grab at the expense of young men and women.

Finding a balance in my persona or divine masculinity has allowed me to embrace saying "I love you" to most men I know. It comes from my experiences with suicide and wondering if anyone on this earth loved me when I couldn't love myself. I want everyone to know that they are loved. Speaking from a shamanic perspective, all illness, both physical and mental, can be summarized as either 1) too much of the wrong energy, 2) not enough of the right energy, or 3) both. Simply put, when our personas come into balance and alignment, remove negative energy (heal the traumas) and add in the positive energy, our being begins to heal from both physical and emotional illness and come to a natural place of homeostasis, Zen, and peace.

YOU ARE THE SHAMAN; YOU ARE THE MEDICINE

Theoretically, you don't need me, or any other shamanic practitioner for that matter. And if you meet someone who says that you do or that only they can cure you, run away quickly. I have often been criticized for being involved in Shamanism as a white, cisgendered, hetero, middle-aged male. There is a lot of gatekeeping in this community—those are simply triggers that those standing at the gates have yet to resolve. Sometimes they will say, "Shamanism has always been rare; shamans were hard to find. It seems that these days, everywhere I look, there is another shaman or healer; this is not the way it was." And they are not entirely wrong. But a couple of things to consider: 1) There is a much larger population of people to serve, so there will be a larger population of healers. 2) Social media makes it possible for us to be aware of what is happening in the world's remote corners. It skews our perception of reality like never before, and 3) We are at a critical juncture in humanity. As a human race, we have never been this sick or had this many souls enrolled in "school." Wouldn't it make sense that souls who have performed that role in past lives would begin flooding into this earthly realm, prepared to assist, teach, heal, and wake up humanity?

"It"

And this is my final core principle: ultimately you are a divine being of light, created from stardust, embodying a physical body made up of recycled butterfly wings and dinosaur bones, reincarnated for what might be the one thousandth time, on a blue crystal that is hurtling through an ever-expanding universe to learn the lessons and nature of universal consciousness to transcend it all—you got this. Sure, someone like me might need to "jump start" the process, but then get busy on your own. The divine creator would not send you here, incapable of learning, healing, and awakening by yourself. The power is within all of us, and we are all equal.

I suppose that if a bastard child, raised in a religious cult, in a small Utah town who grew up to be a Green Beret and found himself engulfed in a 20 year war, who fell down the abyss known as C-PTSD, alcoholism, opioid addiction with a violence arrest record- can claw his way out, learn his place and role in life, the truth of the universe, and mend his relationships, find forgiveness, love and peace- you can too. My invitation to you: do whatever it takes to wake up, to experience "it", to see "it", to prove or disprove it all, to learn your soul contract and account for your karma, and come to that place of knowing, that place of peace, and Zen. To use this "semester" for what it is meant for. That's what this entire experience, this lifetime, this earth, this universe is for.

THE END

APPENDIX OF RESOURCES

BOOKS

- Psychedelics
  - *How to Change Your Mind: What the New Science of Psychedelics Teaches Us About Consciousness, Dying, Addiction, Depression, and Transcendence Hardcover* – May 15, 2018, by Michael Pollan
  - *When the Body Says No: Exploring the Stress-Disease Connection*, Paperback – January 1, 2011, by Gabor Maté, M.D.
  - *The Immortality Key: The Secret History of the Religion with No Name*, Brian C. Muraresku
  - *The Psychedelic Experience: A Manual Based on the Tibetan Book of the Dead– Unabridged*, Paul Heitsch (Narrator), Timothy Leary (Author), Ralph Metzner (Author), Richard Alpert (Author), Daniel Pinchbeck - introduction (Author), Vibrance Press (Publisher)
  - *The Body Keeps the Score: Brain, Mind, and Body in the*

Healing of Trauma, by Scott Pratt, Bessel A. van der Kolk, et al.

o *The Myth of Normal: Trauma, Illness, and Healing in a Toxic Culture,* by Gabor Maté, MD, Daniel Maté, et al.

o *The Psychedelic Explorer's Guide: Safe, Therapeutic, and Sacred Journeys* Audible Logo Audible Audiobook – Unabridged, James Fadiman, PhD (Author), Ross Douglas, Narrator), Inner Traditions Audio (Publisher)

o *When Plants Dream: Ayahuasca, Amazonian Shamanism, and the Global Psychedelic Renaissance–* Unabridged, Daniel Pinchbeck (Author), Sophia Rokhlin (Author), Aaron Shedlock (Narrator), Lauren Ezzo (Narrator), Vibrance Press (Publisher)

o *The Fellowship of the River: A Medical Doctor's Exploration into Traditional Amazonian Plant Medicine–* Unabridged, Joseph Tafur, MD (Author), Joseph Tafur (Narrator, Publisher), Luis Robledo (Narrator)

o *Acid Test: LSD, Ecstasy, and the Power to Heal, Tom Shroder* (Author), Arthur Morey (Narrator), Brilliance Audio (Publisher)

o *Your Soul's Plan"* by Robert Schwartz. This book is one of the most comprehensive resources on soul contracts. Through interviews with mediums and channels, Schwartz outlines how souls plan major life events, such as illnesses, accidents, relationships, and challenges, before incarnation.

o *Journey of Souls* by Michael Newton, Ph.D. A hypnotherapist, Michael Newton, conducted past-life regressions and deep hypnosis sessions with clients to explore life between lives, including the concept of soul contracts.

o *The Seat of the Soul* by Gary Zukav. Zukav's book touches on soul agreements and the purpose of certain

life experiences. He explains that each soul comes into a body with a plan for spiritual growth.
  ○ *Sacred Contracts* by Caroline Myss. Caroline Myss outlines the idea that we make sacred contracts before birth, choosing our archetypes and life experiences for spiritual growth.
  ○ *The Law of One* (Ra Material) In this series of channeled texts, the Law of One teachings elaborate on the idea that souls enter into pre-birth agreements or soul contracts to work through karma and gain spiritual awareness.
  ○ *Between Death and Life* by Dolores Cannon Dolores Cannon, a past-life regression therapist, offers insights from sessions with clients who experienced life between incarnations.
  ○ *Soul Contracts: Find Harmony and Unlock Your Brilliance* by Danielle MacKinnon

- Spirituality
  ○ *Becoming Nobody*: The Essential Ram Dass Collection, Ram Dass (Narrator, Author), Sounds True (Publisher)
  ○ *Zero Limits* Dr. Hew Len's collaboration with Joe Vitale

HOW TO GET IN TOUCH

info@seidrhealing.org

seidrhealing.org

naturalworship.org

@awakegreenberet on social media

www.ingramcontent.com/pod-product-compliance
Lightning Source LLC
Chambersburg PA
CBHW060405130626
46555CB00005B/1992